U0106510

*As the old saying goes,
" life is uncertain,
eat dessert first"*

自 序

~ Making dessert is easy, with direction ~

甜品，她的魅力在於令人心情開朗，是正餐後的重要儀式；沒有了它就像一件事還未完成，心裏依依，念念不忘——她就像性高潮後的輕輕一吻，無論主菜如何激情澎湃，缺了她就像空餘幽幽遺憾。

簡單而平凡不過的食材：麵粉、糖、蛋、奶、牛油，就能做出層出不窮的甜品；加上朱古力、水果、果仁、香料等，不同配方，不同製法，或煮熱，或雪藏，或打入氣泡，或放入焗爐，它會膨脹、鬆軟、香脆、凝固、溶化——它的美味和千變萬化讓人無法抗拒，而一旦迷上它的製作魔法，更是如飲醇醪，不能自拔。

中學時期，我便和朋友在家裏「閉門造餅」，光為了一個忌廉蛋糕，不知耗費了多少盒淡忌廉。1999年，我開始正式學習甜點製作，從港式的牛油蛋糕、海綿蛋糕、忌廉餅，以至有機會現場觀摩世界級甜點師、朱古力師的製作藝術，大大擴闊眼界。自此我開始努力搜羅食材、書籍、向高人請教；在未有DIY店舖和網絡世界的年代，朱古力、榛子醬、果茸等多種食材也得大批購買，為的是向歐洲甜品大師著作裏的極品進發。

法國是甜品巨擘，領導甜點的潮流，外觀、味道都非常考究，跟製作藝術品無異。幼滑、濃郁、鬆脆及清新的口味各領風騷，設計則清麗簡約，不落俗套，裝飾更要以蛋糕的原材料為主題，且全部皆可食用，也不能胡亂配搭，如將咖啡味蛋糕配水果裝飾等等，或蛋糕、奶油不多於四款組合。最新的潮流則數杯裝甜品，現在自助餐桌和高級餐廳也以玻璃櫃展現形形色色的主題甜點。

要吃慣鬆軟蛋糕的香港人接受新的甜點，是漫長的歷程。從牛油忌廉餅進化至鮮忌廉餅，再演變至芝士蛋糕，已經十多年，但似乎就此停滯不前。每每問朋友想吃甚麼餅，他們只會說芝士餅，尤其是Tiramisu的熱潮，至今仍未降溫。研製甜品的先驅國家以法國為首，其次日本、西班牙、比利時、美國、意大利等，近年研究高級朱古力為主的新穎甜點。普遍香港人還未能接受用一頓快餐的價錢來購買一件甜點，可是這件甜點背後的心血、所用的材料、設計者的心思，又怎可衡量呢！

儘管好像遲了一點，不少世界級甜品名店開始進駐香港，近年吹起甜品DIY熱潮，使學習甜品、購買食材都容易多了。我常說，要吃甜點，要不吃最高級的，要不自己動手做，才可吃到真材實料和自己喜歡的甜點。"Live is easy, with Direction"，製作甜點也是一樣。甜點書必須步驟精確，圖文結合，並介紹食材、說明製作竅門等等，更需要獨具匠心的攝影來演繹高雅的作品。本地書籍往往篇幅和製作條件有限；外文書籍固然不乏圖文並茂的專著，但往往礙於翻譯和地域問題，步驟難以揣摩或食材難求，看了光摩拳擦掌，讓人徒喚奈何！

　　走過很多迂迴的冤枉路，我仍堅信"Making dessert is easy, with direction"。希望這本資料全面、深入淺出的甜品書，讓大家在烘焙道上不用再時刻如履薄冰，越來越多甜品愛好者如我，在家中廚房也能做出媲美專業的甜品。

　　這本書的拍攝時間比第一本充裕，拍攝及預備期間全體人員默契十足。特別感謝網友查查的丈夫屎屎和專業的威哥付出的時間和心血，為甜品賦予生命。看到成品照片，腦裏不期然想起拍攝時的快樂片段。但願這份快樂，延展到你的廚房中。

獨角仙

獨角仙的**簡介**

香港烘焙專業協會會員
曾任酒店達六年
業餘烘焙愛好者
http://hk.myblog.yahoo.com/kin-4430
Facebook A/C: http://www.facebook.com/thebluegate

代序

梁樂生＠香港會糕餅主廚

認識阿堅已經有一段日子，從她身上體會到她對烘焙的熱誠和衝勁，今次這本書希望與讀者分享一些新的體驗。

綠萼

透過那一扇玻璃窗，一雙明明很小卻瞪得很大的眼睛，興奮地等待一盤麵包出爐。

如果沒有亞堅，我想我還只是懂得麵包，雪芳，麵包。

第一次上她家，她説：「我很想推廣烘焙，讓更多人懂得做甜點和麵包。」這豪情壯語，不但讓我「霎時感動」，也記憶猶新，只是我沒想過會成了見證人，看着她的Blog一次又一次帶起潮流，然後是第一本書的出版，還開了數次麵包研習班，極一時之盛。

亞堅多年來力臻完美，從沒因為業餘的身份而對食材、工具和知識的追求退讓半分，志氣、勇氣懾人，比她的心靈手巧，更讓我欽佩。可是麵包不過是她的口糧，甜品才是她的拿手好戲與「高潮」所在。有時我想，給她糖油麵粉已經吃得人痴痴迷迷，還惹了一大群瘋子陪着她瘋，如果她是女巫那還得了。數年下來，我由一竅不通，到現在懂得弄點小東西，或許我應尊她為師父，但我半點不願意——永無下山之日，也不會有「青出於藍」這回事，多沒趣。

然而，對於她的甜品書，我實在以弟子的心情屏息期待，那欣慶之情，我想她也不能想像。對於她，可能不過是一次美的追求——把作品拍得像外文書般漂漂亮亮，食譜嘛，她是與朋友分享慣的，只是有精美的書抱在懷裏，感覺真實；對於我，終於盼得平日在亞堅口中説出、很希望——擷取以供諸同好的心得，編成一本所有烘焙愛好者都愛不釋手的書籍，好像終於圓了那眾裏尋書的美夢。

抱歉書籍製作期間公務纏身，未能全力協助，像品嘗了主菜卻苦無甜品般遺憾。惟盼見諒。

羅洪基＠日航酒店糕餅主廚

與獨角仙相識五、六年，最初她在我腦海中的印象只是一位平凡、而且毫不起眼的「師奶仔」而已。然而，透過一次南非之旅後，才發覺在她身上具有不平凡的驚人魅力！獨角仙對有關飲食製作的熱誠絕對是癡中之癡，無論在製作飲食的要求方面、品質方面，甚至工具方面她都相當嚴謹，力求完美、創新，專注投入的精神實在令人佩服。每次走訪其他國家，她必定涉獵當地特色的食材、器具，可見其熱愛程度和認真度極高，說她是「瘋癲」，一點並沒有誇張。

值得欣賞和讚嘆獨角仙既擁有正職之外，仍能把握有限的時間，成就自己獨特的專長，做自己最喜歡、最投入、最擅長的事情。並將技能極致揮灑，創出有聲色的藍色大門，這可不是人人能辦得到的事情呢。再者，能夠在沒有壓力和資源限制的心情下，製作出高質素且受人讚同的糕餅，這是行內人士都未必做到的事。你是一位很不平凡的業餘愛好者，相信你的能力是獨一無二，更是大家羨慕的成功者！你的新書將會帶給更多愛甜品的朋友驚喜與滿足。

曾智波＠香港深灣遊艇會遊艇俱樂部餅師

「比專業更專業」，這是對獨角仙的印象。「藍色大門」讓我明白熱誠便是走向專業的推動力，每星期網誌的新作品、新嘗試、在花園中一手一腳地築起型爐、花上數個寒暑才能閱畢的珍藏烘焙書籍、珍貴的室內空間變成廚房的領土、每一次的假期也花在烹飪……，這份熱誠怎不能叫人敬重和學習。

對前輩的尊重、同輩的友善、後輩的指導，不論是專業從業員還是業餘愛好者，讓她慢慢成為了大家的好友。每一次見到她，總是帶着親切的問候和親手製作的美食，除了朋友關係，慢慢地把她當作我的老師一樣，因為除了技術外，她的品德都是我值得學習的地方。

欣賞過上一次的著作《天然麵包香》後，更加期待這一本甜品書的來臨，因為這會是一本實用、時尚、充滿創意和熱誠的烘焙書籍。

目錄 *Contents*

Entrement

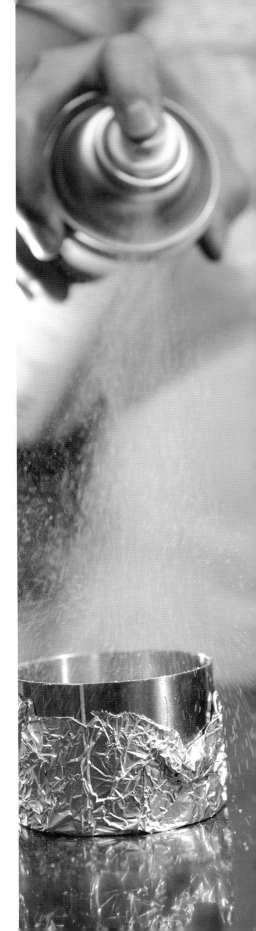

在製作甜點前⋯⋯

熟悉食譜

做甜點前先要熟悉食譜，尤其當中的份量、材料、製作方法，然後按部就班地去製作，千萬別在緊急的關頭去做一些自己未試過做的甜點，也不要試圖做一些自己未熟練的裝飾。

我們要開始做甜點，先要多看幾遍食譜，計出所需份量，列出要購買食材清單，選購正確食材，預備製作地方、器材和用具。其次要注意製作環境的溫度，儲存成品的方法和運送的預備。看似很簡單的東西，如果做不得其法也會事倍功半，更甚者會導致失敗。

購買正確的材料

選購正確的食材是很重要的，不同產地所出產的原材料都有不同的特質，例如歐洲產的麵粉比較日本的磨得粗糙，因此讀者要因應產品的差異去調整份量，不可一概而論。另外，盡量選購和書上相同的食材，才能做出接近的成品。有時候同一樣貨品但不同品牌和價錢，就會有很大出入，以朱古力粉為例，要選購質優的品牌才能做出色澤鮮明、味道濃郁的成品。

預備材料的先後次序

我們要思考預備步驟的先後次序，想過程快捷暢順，就要計劃好工序；例如要先做好須要冷藏的餡心，再預備蛋糕體，接着做其他部份，待所有工序完成，餡心已雪好，這樣就可以節省時間，增加效率。其次是要留意焗爐有沒有預熱，別一開始就開機打麵糊啊！還要留意模具是否已預備好，並塗了油或墊了油紙，擠花袋又是否鑲好擠咀，麵粉有否過篩，這些看似瑣碎的小事，都是做好甜點的重要事項。

正確量度材料

很多食譜尤其是美國或英國食譜書仍採用舊式的度量衡，例如一量杯、一量匙、安士等等；又例如使用彈簧式的磅秤等，可能會因觀看的角度而會有幾克的出入，使用電子磅就能正確地量出所需重量。

選擇適當的器具和器皿

要事半功倍，就要適當地選擇所使用的器具和器皿。以打忌廉為例，要選用比它體積大好幾倍的器皿去盛着攪拌，這樣才不會在攪拌途中忌廉四濺，不但弄污枱面更可能導致份量不足，所以容器大小也會影響製作過程。留意使用是否可入微波爐或入焗爐的器皿，玻璃器皿適宜微波爐使用，但不適合攪拌使用，因玻璃一旦破裂，混入食物中很難看出，十分危險。厚身的不銹鋼大碗適合攪拌用，但不能入微波爐。另外，甚麼時候使用手提打蛋機，甚麼時候只用打蛋器或大型打蛋機，也要因應製作的需要而有不同。

注意攪拌的力度

甜點製作很多時候是運用對材料的攪拌力度或速度去達致所需效果，例如對牛油攪拌的蓬鬆度去製作不同口味的餅皮或曲奇，這點點的變化，要靠大家一點點累積的經驗去分辨的。

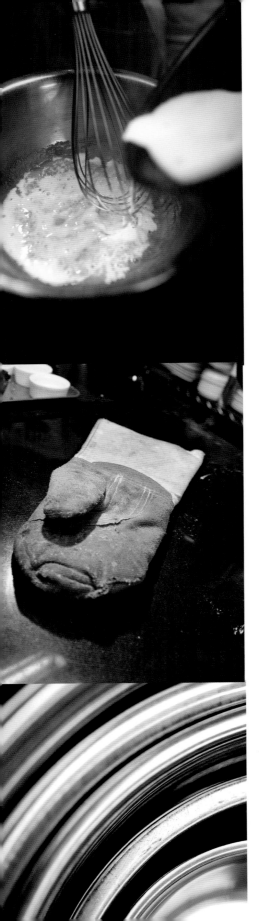

工欲善其事，
必先利其器

焗爐

　　筆者建議在可行情況下買一部容量較大、火力強勁的焗爐。雖然現今一般的廚房不會很大，無法容納一部大型的焗爐，但當你一旦走上烘焙的道路，你就會發現小焗爐的不足。

　　筆者常聽到很多朋友做蛋糕或糕點時做得不好，當中爐溫不當、火力不足、要分數次才能完成等等問題，都是因為缺乏一部理想的焗爐。電焗爐的火力比用氣體的焗爐理想，有上下發熱線和熱風、可調校上下火力的焗爐是基本的條件。相同牌子、相同型號的焗爐因發熱線的偏差，爐火都會有異，所以筆者常說，要掌握自己焗爐的爐火，必須多試。

手動打蛋器

　　不需要很費勁的拌合時，用手動打蛋器就最適合了。

電動打蛋器

　　使用率最多的一件工具，建議購置一部馬力較大的，用來攪拌較多和打發少量材料時用。

座枱電動打蛋機

　　有穩定的攪拌速度和可攪拌較大量的材料；當它在枱上操作時，你可騰出雙手去預備其他材料；要不要購買，視乎你的實際需要。

矽膠蓆 / 矽膠模

　　矽膠蓆和矽膠模在烘焙界的使用率極為普遍，具備大量款式和形狀供選擇，可接受攝氏-40度至+280度溫差，可放入焗爐，但不能接觸明火。可以冷凍，脫模便利，可摺合存放，清潔方便。要注意的是做慕絲或啫喱都要放冰格雪至結冰，成品出模才美麗光滑；如只放雪櫃就會黏着模具，變得不美觀。

市售的矽膠蓆/矽膠模因產地不同而價格各異，法國或意大利的出品比較有保證和耐用。其中專業用的牌子Demarle是筆者最喜愛的品牌，用纖維織網，再注入矽膠，特別薄和傳熱，而且脫模百分百成功，也不需要再在模具上額外噴油，雖然價格較貴但絕對值得投資。但用其他牌子的矽膠模，宜在使用前噴少許油劑在表面，令出模暢順。

金屬慕絲圈 / 模

以不銹鋼和鋁質最為常用，入爐或冷凍皆可，特別的款式可向店舖訂製。

不銹鋼碗

預備不同尺碼的不銹鋼碗，用來盛載材料或用作攪拌器皿都非常合適。

膠刮 / 刮刀

膠刮或刮刀可協助你乾淨地清理黏在碗底或碗邊的殘留物，用來協助攪拌或混合麵糰都很好用。

隔篩 / 粉篩

除了篩麵粉外，篩子還可以用來修飾派撻皮、隔去雜質等等。

擀麵棍

一枝較大的擀麵棍會幫助你擀麵時少費點勁，而且力度亦較平均。

毛掃 / 矽膠掃 / 毛筆

用於蘸糖水塗蛋糕、塗蛋漿、掃蛋糕屑、掃煮糖時的結晶粒、朱古力裝飾上色粉等等。

轉台

切餅、拔餅、擠花、裝飾、噴朱古力時都很有幫助，以金屬製、重身的較理想。

擠花袋

　　用完即棄的擠花袋方便衛生，無論裝飾或入模皆非常有用，是製作甜品的必備用具。

擠花嘴

　　以無焊口的不銹鋼製品較理想，現在亦出產了聚碳酸膠擠嘴，更衛生、更講究，很多高級餅房採用。

電子磅

　　甜點製作差一兩克已很重要，電子磅這精確的量度儀器便是最理想選擇。

溫度計 / 紅外線溫度計

　　用以煮糖、製作海綿蛋糕、朱古力調溫、做意大利蛋白霜時測量溫度。紅外線溫度計和溫度計會有一至兩度的誤差。理想的溫度計可耐溫攝氏-50度至+250度或更高。請選較長的不銹鋼測量針，溫度計便不易因觸碰鍋邊而過熱溶掉。

電動棒狀攪拌器（Bamix）

　　這種被筆者稱為神仙棒的東西，幾乎在任何要攪拌或攪碎的情況也大派用場，例如朱古力調溫、打氣泡、碎蔬果、攪滑等等，絕對是值得投資的工具。

甜點材料簡介

麵粉

　　麵粉是製作糕餅用量最多的基本原料。它由小麥磨製而成，依小麥的的蛋白含量而分為高筋粉、中筋和低筋粉等三種常用的麵粉。高筋粉的蛋白質含量為11.5%-13.5%；中筋粉8.5%；低筋粉蛋白質含量低於8.5%。

　　高筋粉顏色較黃，抓起來乾爽有光澤，用手抓時不易黏成一團，主要用來製作麵包。

　　中筋和低筋粉色澤較白，黏度較低，用手抓很易結成一團，一般用於製作餅乾蛋糕。

　　使用麵粉前必須先過篩，打開包裝後最好保存在密封的容器內，並盡快用完。

糖

　　糖由甘蔗、甜菜頭或玉米提煉出來，種類繁多，有白糖、紅糖、黑糖和葡萄糖等，是乾性糖。糖不但決定糕餅的甜度，還能增加香味、黏性、光澤、保濕、穩定麵糊、防止老化和防腐。糖更有雙重特性，在多水分的配方中能使成品柔軟，但用在水分少的配方就能令成品硬脆。

　　濕性糖有蜂蜜、楓糖、麥芽糖、玉米糖、轉化糖、葡萄糖等，各有用途。

　　本書一般使用細砂糖。

蛋

　　有助於起泡、膨脹、鬆化和增加色澤。蛋的新鮮度對製作甜點影響很大。蛋的用法視乎所製作的甜點而定，就蛋黃來說，越新鮮越理想，但蛋白會因應不同食譜而須要使用冷凍的或室溫的。筆者會在個別食譜詳述。

　　本書使用德國雞蛋，去殼後約重60克，較一般食譜使用的50克較大。其中蛋黃約20克，蛋白約40克。

牛油

　　牛油是牛奶的副製品，分為無鹽牛油和有鹽牛油，歐洲牛油添加了乳酸菌發酵，稱為發酵奶油，香味更強烈，更耐熱。有鹽牛油約有2%鹽分，為了不影響成品的味道，不建議使用。

　　牛油令成品滋潤、光滑、鬆化、香脆，使用優質的牛油令成品口感更濃郁。不同甜點需使用不同溫度質感的牛油，要注意控制牛油的狀態。

　　本書採用法國無鹽牛油。

朱古力

　　朱古力是由可可樹所生長的果實，即可可果製成的食品。從可可果取出果肉內的可可豆，經提煉後製成可可塊、可可粉和可可油。我們常用的三色朱古力（黑、白、奶）是由可可塊和可可油加入不同配方的糖、奶和香料而成。因應不同的甜點和個人喜好，可選擇使用不同品牌、可可含量和風味的朱古力。

　　要注意朱古力不可接觸水分，否則會影響製作，難以令它順滑。要儲存在乾燥清涼的地方，但雪櫃是不適當的。

　　本書多選用比利時65%可可含量的朱古力，取其濃度、甜度、油份適中。

甜 點 副 材 料

忌廉

　　忌廉分淡忌廉和甜忌廉兩種。

　　淡忌廉由牛奶製成，為動物性忌廉，常用的有數種：稀忌廉（single cream）乳脂量約15%-18%，無法打發，有時候又稱為coffee cream 或 table cream，用於沖調飲料。濃忌廉（double cream）乳脂量約36-40%，可以打發，用於製作撻、雪糕和汁醬。攪打忌廉（whipping cream）乳脂含量約35%，可打發。不同品牌有不同的香醇度和顏色，打發體積和耐放程度各異。用於製作冷凍甜點如慕絲、鬆餅、雪糕等等，用途廣泛。

甜忌廉是由奶固體加入植物油如棕櫚油、椰油等，再加糖、乳化劑、水、香料等製成，雖然價錢便宜、儲存時期長，但味道遠較淡忌廉遜色，成份亦非天然，故本書只用於餅面擠花裝飾或拌入大量淡忌廉作抆餅用。

　　淡忌廉在做冷凍糕餅時需要打成七分起泡，軟硬度要適中，打發得太硬，成品就會不夠幼滑。

魚膠

　　魚膠由動物骨膠提煉出來，有片狀和粉末狀兩種。市售的魚膠黏力約在200個單位，所以選擇任何一種都有相同效果。魚膠有吸收水分、凝固物質的作用，口感煙韌，但素食者不宜進食，可用四倍特幼可可油代替。

　　使用粉末魚膠時，先用6-7倍冷水浸發，坐在熱水或在微波爐加熱至完全溶解成清澈液體。片狀魚膠只需將它浸入冷水內待其變軟，撈起後再坐在熱水或在微波爐加熱至完全溶解成清澈液體，或浸軟撈起後加入熱的蛋糊內拌至溶化，十分方便。

　　不論使用粉末魚膠或片狀魚膠，都要注意每次的浸發水分份量。即是說，例如使用魚膠和水的份量是1:7，即1克魚膠用7倍水分，即共8克。雖然浸發後的魚膠未必有7倍，那就須要加水分至8克。每次按這比例做使用魚膠的甜點，就會得出每次相同軟硬度的成品。

香料、甜酒

　　甜點加了香料或甜酒就能提升其味道。

　　雲呢拿豆莢是用量最多、最廣泛的，以馬達加斯加和大溪地出產的最佳，越長越肥大越質優。其他常用的香料有玫瑰、薄荷、肉桂、豆蔻、丁香等等，能適當地配合使用，令甜點味道更多元化。

　　甜酒或餐後酒常用的有rum 酒、kirsch櫻桃酒、蘋果酒、咖啡酒、橙酒、桃酒等等，可加進任何冷熱甜點，按照不同甜點使用不同酒類來增加風味。筆者常備不同味甜酒煮成的糖水在雪櫃，方便製作蛋糕時使用。在蛋糕體上噴少許，可以增加蛋糕香味和濕潤度，比使用人造香料更健康。

鹽

有人會覺得，已用了有鹽牛油，就不用再下鹽吧。實際上，有鹽牛油的鹽分是固定的，不同品牌產地所使用的份量亦有異，所以我們要把這兩種食材分開使用。鹽可調和材料的味道，令甜點不再是單調的甜味，變得有層次感，更可令味道濃縮，做出風味絕佳的成品。筆者喜歡使用法國海鹽，間中使用海鹽花。

冷凍果茸和冷凍水果

大家都知道香港很少出產水果，有的都是產量少，品質不穩定的貨色。要確保每次出品的甜點都有一定水準，外國供應商已研製出由新鮮水果製成的水果果茸和冷凍水果供業界使用。此類冷凍果品質量穩定而且款式繁多，解決了不少找材料的麻煩。雖然我們是住家式的小規模製作，但市面各diy零售商已很有頭腦地把大批的貨品分拆成小包裝出售，令我們也能享用有水準的貨品。

杏仁

香港入口的杏仁主要來自美國和法國，亦有名氣的餅店在西班牙訂購原粒杏仁回來自行加工使用，但較昂貴。杏仁粉是多種法式甜點的材料，很多食譜要將杏仁粉過篩，如此十分浪費，可以把它和糖粉一起放入食物處理器攪碎成粉末狀使用。杏仁產品除了有杏仁粉外，還有杏仁片、杏仁條、杏仁角、杏仁膏、杏仁醬、杏仁糖等等。

榛子

常用的榛子分原粒帶皮、原粒去皮、榛子粉或經加工的榛子醬，無論加進餅底、混入慕絲、製成脆果仁或用作餅面裝飾都是很具風味的食材。唯其味道很濃，在配搭使用時要注意份量，切勿喧賓奪主，遮蓋主材料的風味。

合桃

香港入口合桃的地方有中國、美國和法國。中國合桃無論帶皮或去皮質量都較遜色，美國合桃味淡，有少許回甘和澀味，最優質要數法國合桃，味甜不澀，回甘且果味濃郁，不用烤烘也很好吃。用來配蘋果、無花果、朱古力、香蕉等製作甜點都很美味。

開心果

　　無論原粒開心果、開心果醬都是矜貴的食材，以伊朗出產為上品。這裏並不是指製成小吃的開心果，而是生果仁。筆者喜歡以櫻桃酒去帶出開心果的獨特香味，配上新鮮櫻桃或杏果來製作甜點尤其可口。生開心果果仁很不耐放，在光線照射下容易變黃，要用錫紙袋封好密封儲存。市售開心果醬分為有添加色素、調味醬和天然數款。天然開心果醬約含90%果仁、10%糖，價錢較貴，但味道濃郁。

椰子

　　無論鮮椰子、椰茸、椰子汁、椰奶或椰子粉都帶有東南亞風味，配上不同組合，能創出多樣化的口味。

　　所有果仁或製成品在使用前請檢查清楚有否變壞，因果仁類食品容易吸收其他食物的味道和水氣，較難保鮮。

芒果粉紅胡椒意大利米布丁
Rice Pudding Verrine with Sautéed Mango and Pink Peppercorn

意大利米質感煙韌、有彈性，煮成布丁軟滑可口。雲呢拿籽為意大利米和牛奶賦予靈魂，配以芒果與粉紅胡椒，溫柔與激情同時享有。

製作次序	
1.	意大利米布丁
2.	炒澳洲蘋果芒
3.	裝組

意大利米布丁 ❶
Rice Pudding

材料

意大利米	45 克
牛奶	500 克
砂糖	30 克
蛋黃	40 克
無鹽牛油	20 克
鹽	適量
雲呢拿豆莢	半枝
（用小刀刮出雲呢拿籽）	

 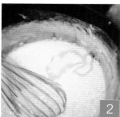 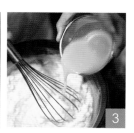

做法

1. 意大利米與牛奶、砂糖及雲呢拿籽同煮至滾。（看圖 1）

2. 轉小火繼續將米煮至軟身及水分收乾至三份一，間中攪拌。蛋黃隔篩過濾雜質，快速倒入鍋裏與牛奶及米混合，煮至滾，拌入鹽及牛油，放涼待用。（看圖 2-3）

炒澳洲蘋果芒
Sautéed Mango

材料

澳洲蘋果芒 ❷	2 個
砂糖	約 100 克
檸檬汁	少許
粉紅胡椒碎 ❸	少許
無鹽牛油	約 40 克

做法

無鹽牛油放煎鍋內，將澳洲蘋果芒和糖一起炒至芒果開始轉深色，下粉紅胡椒碎、檸檬汁調味，關火。(看圖 4-5)

裝組

舀約半杯米布丁進杯內，再放上半杯炒澳洲蘋果芒，放上朱古力裝飾。（看圖 6-8）

貼士

❶ 煮米時要不時攪拌，以免黏住鍋底。

❷ 呂宋芒果雖然香甜，但肉質較軟容易煮爛；請選用半生的澳洲蘋果芒，其果肉較硬，炒時不易散。

❸ 粉紅胡椒色澤嬌艷，配上黃色芒果肉除了能帶出甜品味道，更能增加美感。

Rice Pudding Verrine with Sautéed Mango and Pink Peppercorn

Production workflow

1. Make the rice pudding.
2. Sauté the Australian honey gold mangoes.
3. Assemble.

Rice Pudding ❶

Ingredients

45 g Arborio rice

500 g milk

30 g sugar

40 g egg yolk

20 g unsalted butter

salt

1/2 vanilla pod (split open; seeds scraped out with a knife)

Method

1. Bring rice, milk, sugar and vanilla seeds to the boil. (see picture 1)
2. Turn to low heat. Cook until the rice is soft and the milk reduces to 1/3 of its original volume. Stir occasionally.❶ Sift the egg yolk. Stir the egg yolk in rice mixture. Bring to the boil again. Stir in salt and butter. Let cool for later use. (see pictures 2-3)

Sautéed mango

Ingredients

2 Australian honey gold mangoes ❷ (peeled, cored and sliced)

100 g sugar

lemon juice

crushed pink peppercorns ❸

40 g unsalted butter

Method

Melt the butter in a pan. Put in the mangoes and sugar. Stir until the mangoes start to turn light brown. Add crushed pink peppercorns. Season with lemon juice. Turn off the heat. (see pictures 4-5)

Assembly

Fill half of a champagne flute with the rice pudding. Top with sautéed mango. Garnish with chocolate decorations. (see pictures 6-8)

Tips

❶ Make sure you stir the rice from time to time when you make the rice pudding. Otherwise, it might stick the pot and get burnt.

❷ Filipino mangoes are flavourful and sweet but their flesh is mushy and tends to break down after being heated. I prefer half-ripened Australian mangoes for this recipe because of their firm flesh that tends to stay in one piece after being cooked.

❸ In terms of colour palette, pink peppercorns are in stark contrast with the intense yellow colour of mangoes. Their spiciness also accentuates the true flavour of the mangoes.

希臘風味芝士杯
Arnaglama

意大利鄉村芝士、玫瑰水、合桃、蜂蜜、脆粉粒,迷濛的浪漫與不經意的調皮,重重疊疊化作一杯南歐風情。

製作次序		
1.	脆粉粒	
2.	意大利鄉村芝士忌廉	
3.	裝組	

脆粉粒
Streusel Topping

材料

黃蔗糖	25 克
砂糖	25 克
高筋麵粉	90 克
冷無鹽牛油	50 克
杏仁粉	20 克
肉桂粉	適量
鹽	1 克

做法

1. 將所有材料混和,再以指尖揉碎成粒狀。(看圖 1-2)

2. 放入雪櫃,雪硬後以 180℃ 焗約 15 分鐘至脆身,放涼待用。

貼士

❶ 宜選購 Galbani 的意大利鄉村芝士,芝士糊才會凝結和幼滑。

❷ 玫瑰水和肉桂粉的份量可隨意調節,玫瑰水亦可用玫瑰香油代替。

❸ 益壽糖是一種代糖,由甘露醇和葡萄糖製成。只要用厚底不銹鋼煲將益壽糖煮至微焦,趁熱用匙羹舀在矽膠蓆或不黏布上,待涼就可用來做裝飾。(看圖 3)

意大利鄉村芝士忌廉
Ricotta Cheese Cream

材料

意大利鄉村芝士 ❶	500 克
淡忌廉	250 克
砂糖	100 克
玫瑰花水 ❷	20 克
肉桂粉	3 克

做法

1. 淡忌廉和砂糖打至濃稠。

2. 拌入意大利鄉村芝士、玫瑰花水和肉桂粉,放進雪櫃儲存。

伴食
Garnish and Toppings

合桃(烘香、切碎)	適量
蜂蜜	適量
益壽糖飾 ❸	適量

裝組

杯中倒入一層芝士忌廉、一層脆粉粒,再灑上合桃碎及淋上蜂蜜,然後重複一次,最後放上益壽糖飾。

* 請在食用前才組合。(看圖 4-6)

** 可做約 6 杯

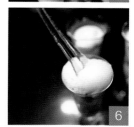

Arnaglama

Production workflow

1. Make the streusel topping.
2. Make the ricotta cheese cream.
3. Assemble in parfait glasses.

Streusel Topping

Ingredients

25 g brown sugar

25 g sugar

90 g bread flour

50 g cold unsalted butter

20 g ground almonds

cinnamon to taste

1 g salt

Method

1. Mix the flour, brown sugar, sugar, butter, cinnamon and salt with tour fingertips. The mixture should be crumbly and should not come together like dough. (see pictures 1-2)
2. Chill until firm and bake in a pre-heated oven at 180°C for about 15 minutes. Set aside until cool for later use.

Ricotta Cheese Cream

Ingredients

500 g ricotta cheese ❶

250 g whipping cream

100 g sugar

20 g rose water ❷

3 g ground cinnamon

Method

1. Whip the cream and sugar until thick.
2. Add the ricotta cheese, rose water and ground cinnamon. Keep it in a fridge.

Garnish and toppings

walnuts (baked and chopped)

honey

isomalt ❸

Assembly

Put a layer of ricotta cream on the bottom of a parfait glass. Top with a layer of streusel, baked walnuts and honey. Repeat this step to build alternate layers. Garnish with Isomalt. Serve. (see pictures 4-6)

** Assemble only before serve.

** Makes 6 glasses.

Tips

❶ For the best result use ricotta cheese from the Galbani brand. The cheese cream tends to set much better and tastes smoother.

❷ You may adjust the quantity of rose water and cinnamon according to your own taste. You may also use rose essence instead of rose water.

❸ Isomalt is an artificial sweetener made with mannitol and glucose. To apply, just melt it in a non-stick pan and cook until mildly caramelized. Then drip or spoon it over non-stick baking lining or silicone mat. Leave it to cool and decorate the Arnaglama with it. (see picture 3)

酒漬覆盆子開心果慕絲杯
Pistachio Mousse with Marinated Raspberries Verrine

杯裝甜品大行其道，配搭繁多如萬花筒，由簡簡單單的軟滑慕絲到層層不同的軟硬酥脆都可以在一杯子內盡顯現。它的層次感和色彩讓人垂涎。

覆盆子、開心果是經典組合。除了鮮覆盆子或冷凍覆盆子外，再加入糖水覆盆子和覆盆子酒，以增加漿果香。

1. 酒漬覆盆子果茸
2. 開心果慕絲
3. 裝組

酒漬覆盆子果茸
Marinated Raspberries Coulis

材料

冷凍覆盆子果茸	300 克
市售樽裝糖水覆盆子 ❶	50 克
覆盆子酒	20 克

做法

1. 將所有材料混合，醃漬 15 分鐘。（看圖 1）

2. 然後攪拌成果泥狀，隔去部分覆盆子籽。（看圖 2-3）

開心果慕絲

Pistachio Mousse

材料

牛奶	125 克
開心果醬 ❷	50 克
雲呢拿豆莢	半條（用小刀刮出雲呢拿籽）
蛋黃	40 克
砂糖	25 克
魚膠片 (用凍水浸軟)	2 克
車厘子酒	4 克
打起淡忌廉	125 克

做法

1. 牛奶、雲呢拿籽一起煮滾。

2. 蛋黃及砂糖拌勻，一邊攪拌一邊慢慢倒入（1），過篩後回鍋煮至 85℃，期間不斷攪拌，再過篩。

3. 加入開心果醬、車厘子酒和已浸軟的魚膠片，拌勻。

4. 放涼至約 25℃，然後拌入打起淡忌廉，放入雪櫃冷藏至凝固。

裝組❸

舀 1/3 杯覆盆子醬在杯底，擠上開心果慕絲至九分滿，放上糖水覆盆子和食用金箔裝飾。（看圖 4-6）

** 可做 6 小杯

貼士

❶ 糖水覆盆子可在大型超市如 city'super 有售，亦可以酒漬覆盆子代替，這時可省去覆盆子酒。

❷ 宜選用原味無加色素的開心果醬，雖然價錢貴一點，但天然的果仁味和人工添加的東西是無可比擬的。

❸ 開心果慕絲放入擠花袋前不要再次攪滑它，這樣才可出現層次感，否則只會變成一堆軟糊，當綴上糖水覆盆子時，就無法做到應有的層次感了。

Pistachio Mousse with Marinated Raspberries Verrine

Production workflow

1. Make the marinated raspberries coulis
2. Make the pistachio mousse
3. Assemble in glasses.

Marinated Raspberries Coulis

Ingredients

300 g frozen raspberries
50 g raspberries in syrup ❶
20 g raspberry liqueur

Method

1. Mix all ingredients together. Leave them for 15 minutes.(see picture 1)
2. Stir into puree. Strain away some of the seeds. (see pictures 2-3)

Pistachio Mousse

Ingredients

125 g milk
50 g pistachio paste ❷
1/2 vanilla pod (split open; seeds scraped off with a knife)
40 g egg yolk
25 g sugar
2 g gelatine leaf (soaked in water until soft; drained)
4 g Kirsch (cherry liqueur)
125 g whipped cream

Method

1. Boil milk and vanilla seeds in a pot.
2. Mix egg yolk and sugar in a bowl. Pour the hot milk mixture into the egg yolk mixture. Sift and pour back into the pot. Heat it up to 85°C. Stir continuously during the cooking process. Sift again.
3. Add pistachio paste, kirsch and gelatine leaf. Mix well.
4. Leave it to cool to 25°C. Fold in whipped cream. Refrigerate until set.

Assembly ❸

Fill the serving glass up to 1/3 of its height with marinated raspberries. Pipe pistachio mousse on top up to 90% full. Arrange raspberries in syrup and edible gold leaf on top. Serve. (see pictures 4-6)

** Makes 6 verrines.

Tips

❶ You can get raspberries in syrup from major supermarkets such as city'super. You may also use raspberries steeped in alcohol instead, in which case you should skip the raspberry liqueur.

❷ Pick unsweetened pistachio paste without artificial colouring. Although it is more expensive, its natural nutty flavour is beyond any manmade flavouring can compare.

❸ Before you put the pistachio mousse into a piping bag, do not beat it with an electric mixer again. Otherwise, it may be too runny in consistency. When you layer it with the marinated raspberries, they might blend in with each other.

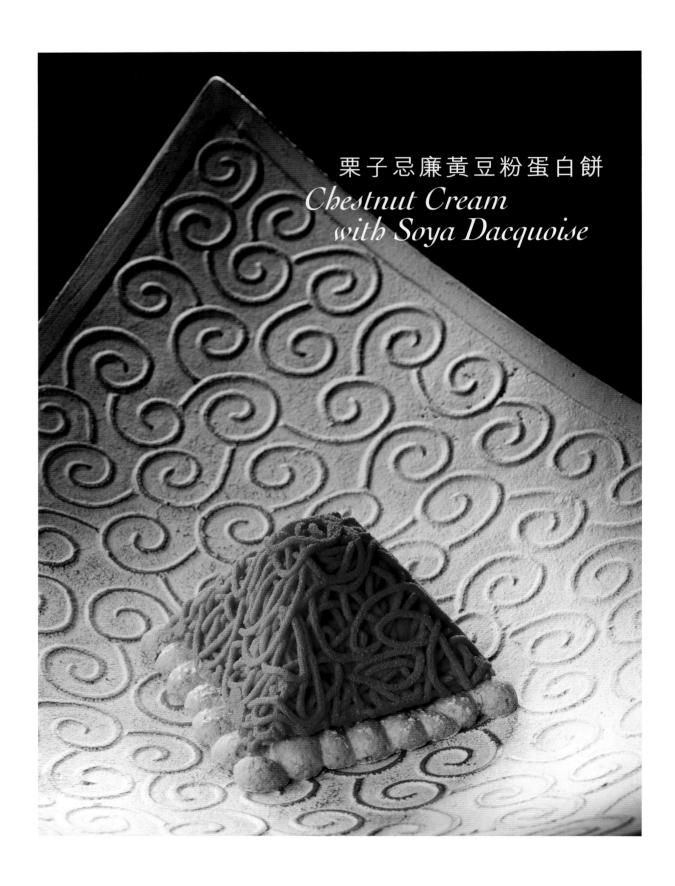

栗子忌廉黃豆粉蛋白餅

Chestnut Cream
with Soya Dacquoise

以栗子為主角的甜品，最經典當數 Mont Blanc；這款以牛奶朱古力、黃豆粉、櫻桃酒漬車厘子搭配栗子的甜點，卻充滿亞洲風情，讓人聯想起秋天的富士山。

製作次序		
1.	栗子忌廉	4. 噴槍用紅色朱古力
2.	黃豆粉蛋白餅	5. 裝組
3.	牛奶朱古力忌廉	

栗子忌廉
Chestnut Cream

材料

HERO 牌罐頭甜栗子茸	300 克
淡忌廉	75 克
蛋黃	40 克
砂糖	6 克
魚膠片（用凍水浸軟）	4 克
冧酒	10 克

做法

1. 蛋黃與砂糖拌勻。

2. 淡忌廉煮滾，倒進 (1) 內拌勻。

3. 將 (2) 過篩，然後倒進鍋內，一邊攪拌一邊加熱至 85℃，使之濃稠。

4. 加入栗子茸、已浸軟的魚膠片及冧酒。需要時可用電動棒狀攪拌器將其攪滑。將栗子糊舀進已套上 2 號圓形擠花嘴的擠花袋內，擠出不規則花紋於金字塔形矽膠模內，並放冰格冷凍至結冰。（看圖 1-4）

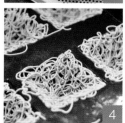

黃豆粉蛋白餅
Soya Dacquoise

材料

蛋白	120 克
砂糖	60 克
白醋	3-4 克
杏仁粉	30 克
糖霜	40 克
黃豆粉	40 克

做法

1. 糖霜、黃豆粉及杏仁粉混合。

2. 蛋白及砂糖攪打至起泡變白，加入白醋，再攪打至企身。

3. 輕輕拌入（1），拌勻成麵糊。(看圖 5)

4. 以擠花袋將麵糊擠在焗盤上（因為這是用作栗子黃豆粉蛋白餅的底部，所以宜比金字塔形矽膠模大一點），篩上糖霜。放入已預熱 160℃ 之焗爐，焗約 25 分鐘至脆身。(看圖 6-8)

牛奶朱古力忌廉
Milk Chocolate Chantilly Cream

材料

牛奶朱古力 ❶	70 克
打起淡忌廉	220 克
櫻桃酒漬車厘子 ❷	100 克

做法

1. 朱古力用熱水坐溶，間中攪拌，直至朱古力溫度達 50℃，立即混入淡忌廉內，快速拌勻。

2. 拌入酒漬車厘子。

噴槍用紅色朱古力
Red Chocolate Spray

材料

白朱古力	200 克
谷咕油	220 克
油溶性紅色粉	適量

做法

將所有材料隔水或在微波爐慢慢加熱溶化至約 50℃，過濾。

裝組

1. 把牛奶朱古力忌廉釀入已冷凍至結冰的栗子忌廉內。（看圖 9-10）

2. 再放入冰格冷凍。脱模，噴上一層紅色朱古力，放在一片黃豆粉蛋白餅上即可。（看圖 11）

** 可製約 10 件餅

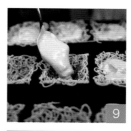

貼士

① 用牛奶朱古力的原因是它味道溫和，不會搶去栗子的味道。

② 櫻桃酒漬車厘子和罐裝櫻桃酒漬車厘子餡料是兩種不同的食材，櫻桃酒漬車厘子酒味更濃郁，品質更矜貴，用來做倒模朱古力的餡心或是用來配雞尾酒食用都一流。

Chestnut Cream with Soya Dacquoise

Production workflow

1. Make the chestnut cream.
2. Make the milk chocolate Chantilly cream.
3. Make the soya Dacquoise.
4. Make the red chocolate spray.
5. Assemble the cake.

Chestnut Cream

Ingredients

300 g canned chestnut puree (Hero brand)

75 g whipping cream

40 g egg yolk

6 g sugar

4 g gelatine leaf (soak in water until soft, drained)

10 g rum

Method

1. Mix egg yolk with sugar.
2. Bring the cream to the boil and pour it into egg yolk mixture from step 1.
3. Sift the mixture and continue to cook at 85°C until it thickens. Stir continuously.
4. Add chestnut puree, soaked gelatine leaf and rum. Use a hand blender to mix well if needed. Put the mixture into a piping bag. Pipe the mixture randomly through a No.2 nozzle to form a layer on the four sides of a pyramid-shaped silicone mould, leaving room at the centre. Freeze it. (see pictures 1-4)

Milk Chocolate Chantilly Cream

Ingredients

70 g milk chocolate

220 g whipped cream

100 g chopped Griottine cherries ❶

Method

1. Melt the chocolate over a pot of simmering water. Stir occasionally until it reaches 50°C. Pour in whipped cream and whisk quickly. ❷
2. Stir in the cherries.

Soya Dacquoise

Ingredients

120 g egg whites

60 g sugar

3-4 g vinegar

30 g ground almonds

40 g icing sugar

40 g soya powder

Method

1. Mix icing sugar, soya powder and ground almonds together.
2. Beat egg whites and sugar until light and fluffy. Add vinegar and continue to whip until stiff.
3. Fold in soya powder mixture from step 1. Mix well. (see picture 5)
4. Pipe the meringue from step 3 onto a lined or non-stick baking tray. As this Dacquoise forms the base of the pyramid cake, you may pipe a square slightly larger than your pyramidal silicone mould. Sprinkle icing sugar over the meringue. Bake in a preheated oven at 160°C for about 25 minutes or until golden and crispy. (see pictures 6-8)

Red Chocolate Spray

Ingredients

200 g white chocolate
220 g cocoa butter
oil-based red colouring

Method

Melt all ingredients over a hot water bath or in a microwave to about 50°C. Pass through a sieve.

Assembly

1. Fill the pyramidal mould with Milk Chocolate Chantilly Cream over the frozen chestnut cream. (see pictures 9-10)

2. Keep in freezer until frozen. Unmould the frozen chestnut cream pyramids. Spray on red chocolate with a chocolate sprayer. (see picture 11)Transfer onto a piece of soya Dacquoise. Serve.

** Makes 10 cakes.

Tips

❶ Griottine cherries and canned cherry pie filling in kirsch are two different ingredients. The former have a richer liquor flavour and are much higher in quality (which also explains why they are more expensive than the latter). You may use Griottine cherries as the filling for chocolate, or serve them straight with cocktails.

❷ Milk chocolate is used in the Chantilly cream because dark chocolate would taste too strong. It might cover up the delicate taste of the chestnut.

菠蘿脆餅
Pineapple Satellite

菠蘿忌廉、椰子啫喱加上脆脆底，熱帶風情洋溢於紙上。

製作次序

1. 椰子啫喱
2. 菠蘿忌廉
3. 白朱力椰子菠蘿脆脆
4. 黃色淋面

椰子啫喱
Coconut Cream Filling

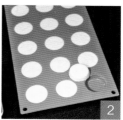

材料

淡忌廉	55 克
砂糖	20 克
椰汁	55 克
魚膠片（用凍水浸軟）	2 克

做法

1. 淡忌廉、椰汁及砂糖一同煮滾。

2. 加入已浸軟的魚膠片，拌勻，注入小圓模內，放入冰格冷凍至結冰，待用。（看圖 1-2）

菠蘿忌廉
Pineapple Mousseline Cream ❶

材料

冷凍菠蘿果茸	250 克
蛋黃	40 克
砂糖	60 克
粟粉	20 克
無鹽牛油 (軟化)(A)	60 克
無鹽牛油 (軟化)(B)	60 克
打起淡忌廉	135 克

做法

1. 菠蘿果茸煮滾。❷

2. 調勻蛋黃、粟粉及砂糖。將已煮滾的菠蘿果茸倒進蛋黃漿裏，過篩後回鍋以細火煮滾，期間不斷攪拌，待涼。

3. 拌入軟牛油 (A)，放入雪櫃待涼，備用。

4. 拌入軟牛油 (B) 和打起淡忌廉。

白朱力椰子菠蘿脆脆
Coconut Crunchy Base

材料

脆脆片	25 克
卜卜米	25 克
白朱古力	50 克
可可油	15 克
烘香椰絲	10 克
菠蘿乾	15 克

做法

1. 白朱古力及可可油一起溶化。（看圖 3）

2. 加入脆脆片、卜卜米、烘香椰絲及菠蘿乾拌勻。（看圖 4）

3. 將（2）壓在餅模底部，冷凍至朱古力凝固。（看圖 5）

黃色淋面 ❸
Yellow Coating

材料

透明果膠	500 克
水	50 克
黃色食用色素	適量

做法

1. 透明果膠與水一同隔熱水加熱至 50℃，待果膠開始溶解，加入食用色素。

2. 用電動棒狀攪拌器慢速打滑，保持約 50℃，待用。

裝組

1. 在模內擠 1/3 菠蘿忌廉，把椰子啫喱釀在中心，再填滿菠蘿忌廉，冷凍至結冰。（看圖 6-8）

2. 把保持在 50℃ 的黃色淋面淋在已結冰的蛋糕上，放在白朱古力椰子菠蘿脆脆上。（看圖 9-10）

** 約可製 8 件菠蘿脆餅

貼士

1 Mousseline cream（La Crème Mousseline）可以意譯成「冒泡的奶油」，以 pastry cream 作基礎，加入少量牛油和淡忌廉令其更幼滑，口感介乎 pastry cream 與 butter cream 之間，質感幼細絲滑。

2 製作菠蘿 mousseline cream 宜使用冷凍菠蘿果茸，如用新鮮菠蘿汁，菠蘿的酵素會令奶油內的忌廉不能凝固，甚至油水分離。如用鮮菠蘿茸，要把它煮至水分收乾少許，再放涼才使用。

3 黃色淋面除了要保持在 50℃ 外，還要一次過大量地淋下，才可以覆蓋全件菠蘿脆餅；如中途停止了，很可能有部分會上不到淋面，如稍後再淋上，就會有厚薄不均的情況出現。

Pineapple Satellite

Production workflow

1. Make the coconut cream filling.
2. Make the pineapple mousseline cream.
3. Make the coconut crunchy base.
4. Make the yellow coating.
5. Assemble.

Coconut Cream Filling

Ingredients

55 g whipping cream

20 g sugar

55 g coconut milk

2 g gelatine leaf (soaked in water until soft; drained)

Method

1. Bring cream, coconut milk and sugar to the boil.
2. Add soaked gelatine leaf, stir until it dissolves. Pour in the round mould and freeze for later use. (see picture 1-2)

Pineapple Mousseline Cream ❶

Ingredients

250 g frozen pineapple puree

40 g egg yolk

60 g sugar

20 g cornflour

60 g unsalted butter (A) (at room temperature)

60 g unsalted butter (B) (at room temperature)

135 g whipped cream

Method

1. Boil the pineapple puree in a pot. ❷
2. In a bowl, mix together egg yolk, sugar and cornflour. Pour the hot pineapple puree into the egg yolk mixture. Sieve and pour back into the pot. Cook over low heat until it boils. Stir continuously in the cooking process. Leave it to cool.
3. Stir in butter (A). Refrigerate until cold.
4. Stir in butter (B) and whipped cream. Set aside.

Coconut Crunchy Base

Ingredients

25 g crunchy wheat flakes (Pailleté Feuilletine)

25 g puff rice

50 g white chocolate

15 g cocoa butter

10 g toasted desiccated coconut

15 g dried pineapple

Method

1. Melt the chocolate and cocoa butter. (see picture 3)
2. Add crunchy wheat flakes, puff rice, desiccated coconut and dried pineapple. Mix well. (see picture 4)
3. Press the above mixture on the bottom of a cake mould. Refrigerate until chocolate has set. (see picture 5)

Yellow Coating ❸

Ingredients

500 g neutral nappage

50 g water

a few drops yellow food colouring

Method

1. Heat neutral nappage and water over a hot water bath to 50°C. When the neutral nappage dissolves, add yellow food colouring.
2. Blend with a hand blender over low speed until the mixture is smooth. Keep the temperature at 50°C for later use.

Assembly

1. Fill the silicone mould up to 1/3 of its depth with pineapple mousseline cream. Place the frozen coconut cream filling at the centre. Then fill the mould up with remaining pineapple mousseline cream. Keep in the freezer until frozen. (see pictures 6-8)
2. Pour the yellow coating at 50°C over the frozen cake. Transfer on top of a piece of coconut crunchy base. Serve. (see pictures 9-10)

** Makes 8 Pineapple Satellites.

Tips

❶ Mousseline cream (or La Crème Mousseline in French) can be roughly translated into bubbly cream in English. It is basically a pastry cream with a little more butter and whipping cream for a finer texture. It tastes half way between pastry cream and butter cream. It is smooth and delicate.

❷ When you make the pineapple mousseline cream, try to use frozen pineapple puree as much as possible. Fresh pineapple juice is not an alternative because the enzymes in the juice would prevent the cream from clotting, so that the mousse will separate. On the other hand, you may use fresh pineapple puree provided that you simmer it down to make it more concentrated and leave it to cool completely before use.

❸ For the perfect coating, make sure the yellow coating mixture is at 50°C. Also make sure you pour enough coating mixture on it in one go to cover the whole cake. If you stop pouring half way and pour again, the coating will be uneven.

咖啡慕絲
Coffee Mousse Charlotte

嘗一口，品味多重、滋味大不同的咖啡香。

製作次序	1. 咖啡凍	4. 咖啡牛奶朱古力慕絲
	2. 手指餅	5. 咖啡焦糖淋面
	3. 咖啡脆片	6. 裝組

咖啡凍
Coffee Jelly

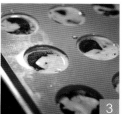

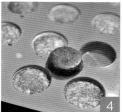

材料

熱水	120 克
砂糖	20 克
特濃即溶咖啡粉 ❶	6 克
魚膠片（用凍水浸軟）	5 克

做法

1. 調勻砂糖、特濃即溶咖啡粉及熱水。（看圖 1）

2. 趁熱加入已浸軟的魚膠片，拌勻，然後注入小圓模內，放入冰格冷凍至結冰。（看圖 2-4）

手指餅
Ladyfingers

材料

粟粉	40 克	蛋白	100 克
砂糖 (A)	30 克	砂糖 (C)	30 克
蛋黃	60 克	高筋麵粉（篩勻）	55 克
砂糖 (B)	30 克	糖霜	適量

做法

1. 粟粉、砂糖（A）混合。

2. 蛋黃和砂糖（B）打至淡黃色和濃稠。

3. 打發蛋白。先以電動攪拌機開至最快速攪打至蛋白起粗泡，約 30 秒，然後分兩至三次加入砂糖（C），蛋白此時開始濃稠。

4. 分兩至三次加入（1），繼續打至蛋白變得很有韌力和非常堅挺。輕輕拌入蛋黃糊和高筋麵粉，用膠刮拌勻成麵糊。

5. 麵糊舀進擠花袋內，然後擠 6 個直徑 4 厘米的圓餅在已墊不黏布或矽膠蓆的焗盤上，篩上糖霜❷。（看圖 5）

6. 放入已預熱 170℃ 的焗爐內，焗約 15 分鐘至金黃色。（看圖 6）

咖啡脆片（裝飾用）
Coffee Brittle

材料

牛油	50 克
黃糖	35 克
葡萄糖膠	40 克
低筋麵粉	20 克
杏仁（切碎）	40 克
即溶咖啡粉	10 克

做法

1. 將所有材料調勻，然後放於兩片矽膠蓆中，擀薄，放入雪櫃冷藏至硬身。

2. 用已預熱 160℃ 的焗爐焗約 12 分鐘。在咖啡脆片還暖的時候，用刀切成所需形狀。

咖啡牛奶朱古力慕絲❸❹
Mocha Mousse

材料

牛奶朱古力（切碎）	100 克	特濃即溶咖啡粉	16 克
淡忌廉	88 克	魚膠片（用凍水浸軟）	7 克
牛奶	88 克	打起淡忌廉	225 克
砂糖	10 克	咖啡酒	8 克
蛋黃	50 克		

做法

1. 調勻蛋黃及砂糖。

2. 淡忌廉、特濃即溶咖啡粉和牛奶一起煮滾。（看圖 7）

3. 把（2）倒進（1），拌勻，過篩，加熱至85℃。（看圖 8）

4. 將（3）直接篩入牛奶朱古力上，拌勻，待溶化後立即加入已浸軟的魚膠片，拌勻，待至完全涼透。（看圖 9-10）

5. 分數次拌入打起淡忌廉。（看圖 11-12）

咖啡焦糖淋面
Caramel Glazing

材料

葡萄糖膠	120 克
砂糖	200 克
牛油清 ❺	60 克
杏桃鏡面果膠 ❻	320 克
暖水	70 克
魚膠片（用凍水浸軟）	8 克

做法

1. 葡萄糖膠、砂糖和水一同煮成焦糖。（看圖 13）

2. 加入牛油清、暖水和杏桃鏡面果膠煮至滾，加入已浸軟的魚膠片，拌勻，再放涼至 30℃待用。（看圖 14）

裝組

1. 在模內擠入 1/3 朱古力慕絲，放上一片手指餅，再放上咖啡凍。（看圖 15-16）

2. 注入朱古力慕絲填滿餅模，冷凍至結冰。將朱古力慕絲移離餅模，淋上咖啡焦糖淋面，後放上咖啡脆片裝飾便可。（看圖 17-18）

** 可做 8 個 7.8 厘米高 x 4 厘米寬的咖啡慕絲蛋糕

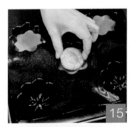

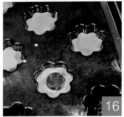

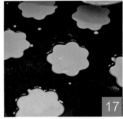

貼士

1 宜選用雀巢的特濃即溶咖啡粉或雀巢金牌即溶咖啡粉。當然，咖啡味的濃淡可自行調節。

2 手指餅入爐前篩上糖霜是非常重要的，它能令手指餅保持形狀，冷卻後不會收縮。

3 控制溫度乃製作慕絲、凍甜點的重要一環。曾加熱的材料必須待完全涼透方可與冷的打起忌廉混合，混合時先加進小部分忌廉，然後再分次加入，讓兩者溫差慢慢接近。如未完全涼透即加入冷的忌廉，魚膠會凝固成小塊，又或者熱溶了打起的忌廉，慕絲變得過稀，失去輕盈絲滑的口感。

4 咖啡牛奶朱古力慕絲要冷凍至結冰，才會令鏡面在倒下時立即凝固，否則鏡面會過薄而看到蛋糕體。所以必須待慕絲結冰才做淋面。

5 牛油清即是把牛油煮溶，去掉渾濁部分後的清澈液體。

6 杏桃鏡面膠可在各 DIY 店有售。

Coffee Mousse Charlotte

Production workflow

1. Make the coffee jelly.
2. Make the ladyfingers.
3. Make the coffee brittle.
4. Make the mocha mousse.
5. Make the caramel glazing.
6. Assemble the cakes.

Coffee Jelly
Ingredients

120 g hot water

20 g sugar

6 g coffee powder (espresso) ❶

5 g gelatine leaf (soak in cold water until soft, drained)

Method

1. Mix hot water, coffee powder and sugar together. (see picture 1)
2. Put in gelatine leaf. Stir well and let cool. Freeze it. (see pictures 2-4)

Ladyfingers
Ingredients

40 g cornflour

30 g sugar (A)

60 g egg yolk

30 g sugar (B)

100 g egg whites

30 g sugar (C)

55 g bread flour (sieved)

icing sugar

Method

1. Mix cornflour and sugar (A) together.
2. Beat egg yolk and sugar (B) until pale and thick.
3. Beat the egg whites over high speed with an electric mixer for about 30 seconds until soft peaks form. Then add half or one-third of sugar (C) at a time. Beat after each addition. Firm peaks should start to form.
4. Then add half or one-third of the cornflour mixture at a time from step 1. Continue beating after each addition until the egg whites are stiff. Gently fold in egg yolk mixture and bread flour with a spatula.
5. Transfer the batter into a piping bag. Pipe six round patties, each about 4 cm in diameter, on a silicone mat or non-stick baking lining. Sprinkle icing sugar on top. ❷ (see picture 5)
6. Bake in a preheated oven at 170°C for 15 minutes until golden. Set aside to let cool. (see picture 6)

Coffee brittle (as garnish)
Ingredients

50 g butter

35 g brown sugar

40 g glucose syrup

20 g cake flour

40 g chopped almonds

10 g instant coffee powder

Method

1. Combine all ingredients together. Spread the mixture between two silicone mats. Roll the mixture out with a rolling pin. Refrigerate until firm.
2. Bake in a preheated oven at 160°C for about 12 minutes. Cut to desired shapes while still warm.

Mocha Mousse ❸ ❹
Ingredients

100 g milk chocolate (chopped)

88 g whipping cream

88 g milk

10 g Sugar

50 g egg yolk

16 g espresso coffee powder

7 g gelatine leaf (soaked in cold water until soft, drained)

225 g whipped cream

8 g Kahlua (coffee liqueur)

Method

1. Mix egg yolk and sugar together.

2. In a pot, pour in whipping cream, espresso coffee powder and milk. Bring to the boil. (see picture 7)

3. Pour the hot coffee mixture into the egg yolk mixture from step 1. Stir well and pass through a sieve. Heat it up to 85°C. (see picture 8)

4. Pour the resulting mixture into the chopped chocolate. Mix well and stir until chocolate melts. Put in the gelatine leaf and stir until it dissolves. Leave it to cool completely. (see pictures 9-10)

5. Stir in some of the whipped cream at one time. Fold well after each addition until well incorporated. (see pictures 11-12)

Caramel Glazing

Ingredients

120 g glucose syrup

200 g sugar

60 g clarified butter ❺

320 g apricot glaze ❻

70 g warm water

8 g gelatine leaf (soaked in cold water until soft; drained)

Method

1. Cook glucose syrup, sugar and water until caramelized. (see picture 13)

2. Add clarified butter, warm water and apricot glaze. Cook until it boils. Add soaked gelatine leaf. Mix well. Let it cool to 30°C. (see picture 14)

Assembly

1. Pipe 1/3 of the mocha mousse into the charlotte mould. Arrange a layer of ladyfinger on top. Put on a slice of coffee jelly. (see pictures 15-16)

2. Fill the mould up with the remaining mocha mousse. Freeze it. Turn the mousse out. Pour caramel glazing evenly on top. Garnish with coffee brittle. Serve. (see pictures 17-18)

** Makes 8 coffee mousse cakes, measuring 7.8 cm x 4 cm each.

Tips

❶ I prefer Nestle instant espresso and Gold coffee powder. Of course, you may adjust the strength of the coffee according to your own taste.

❷ Make sure you sieve some icing sugar over the ladyfingers before baking them. The icing sugar keeps the shapes of the ladyfingers so that they won't shrink after cooling.

❸ The control of temperature throughout the mixing process is an important skill when making mousse or chilled dessert. Any ingredients that have been heated must have cooled off completely before being mixed with whipped cream. Also try to temper the once heated ingredient first by mixing in a small amount of whipped cream into that ingredient. This step brings the temperature of the two ingredients closer. Then pour a little of the mixture back into the remaining whipped cream at one time. Fold well after each addition. If warm mixture is poured right into the cool whipped cream, gelatine in the mixture will form lumps right away. The heat may also melt the whipped cream making the mousse too watery and thin, ruining its supposedly fluffy and smooth texture.

❹ The mocha mousse has to be frozen, so that the glazing set right away when poured over. Otherwise, the glazing will be too thin and the mousse shows through.

❺ Clarified butter is made from butter that has been melted and left to separate. The top clear layer is skimmed off and it is the clarified butter.

❻ Apricot glaze is available from DIY baking supply stores.

兩顆心
Two Hearts

用杏仁膏製作的杏仁蛋糕，香、味極其華麗，非同凡響。把它做成心形，
配合覆盆子牛油忌廉，成為一份華貴的心意。

製作次序	1. 杏仁蛋糕	4. 忌廉
	2. 覆盆子牛油忌廉	5. 裝組
	3. 覆盆子糖漿	

杏仁蛋糕 ❶
Almond Cake

材料

杏仁膏（50%）❷	250 克
全蛋	20 克
蛋黃	40 克
蛋白	70 克
砂糖	5 克
低筋麵粉	25 克
粟粉	25 克
車厘子酒	25 克
無鹽牛油（溶化保溫）	60 克

做法

1. 模具噴上油劑，麵粉、粟粉一起過篩。

2. 全蛋、蛋黃及杏仁膏用電動棒狀攪拌器攪滑，並至奶白色。

3. 蛋白、砂糖打起，將蛋白糊輕輕拌入 (2) 內。

4. 拌入麵粉和車厘子酒，加入溶牛油，拌勻。

5. 舀入模中至八成滿，需要時用小抹刀抹平，蓋上不黏布或矽膠蓆，再放上焗盤，令蛋糕發起時頂部平滑。（看圖 1-2）

6. 放入已預熱 180℃ 的焗爐，焗約 20 分鐘至金黃色，待涼備用。（看圖 3）

  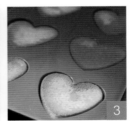

覆盆子牛油忌廉
Raspberry Butter Cream

材料

覆盆子果茸	200 克
砂糖 (A)	250 克
雞蛋	120 克
砂糖 (B)	30 克
軟牛油	300 克
雲呢拿豆莢	1 枝（用小刀刮出雲呢拿籽）

做法

1. 覆盆子果茸和砂糖（A）煮滾。

2. 雞蛋及砂糖（B）拌勻，倒入已煮滾的果茸，拌勻。（看圖 4）

3. 隔熱水將（2）加熱，期間不斷攪拌，直至溫度達 85℃，隔去雜質。（看圖 5）

4. 待果茸蛋糊冷卻後，拌入軟牛油打勻，最後拌入雲呢拿籽。（看圖 6-7）

覆盆子糖漿
Raspberry Syrup

材料

覆盆子果茸	100 克
砂糖	60 克
水	70 克

做法

將水、覆盆子果茸、砂糖煮滾，放涼待用。

忌廉
Chantilly Cream

材料

淡忌廉	100 克
砂糖	20 克

做法

混合所有材料，打至企身，待用。

貼士

1 這蛋糕質感像牛油蛋糕，蛋白加入蛋黃糊後攪拌至蛋白消失，蛋糕焗後才不會像海綿般鬆身。

2 杏仁膏（Raw marzipan）是杏仁的製品，宜購買杏仁成份超過 50% 的製品，因為其餘份量就是糖，杏仁含量越高越香濃。市面上標明用作裝飾的杏仁膏，通常只有 20-25% 杏仁含量，色澤較白，也甜得多，不宜用來製作蛋糕體。

裝組

1. 將蛋糕蘸上覆盆子糖水。（看圖 8）

2. 擠上覆盆子牛油忌廉和忌廉，放上鮮覆盆子和金箔作裝飾。（看圖 9-10）

** 可做 9 件 5 厘米 x 6 厘米 x 3 厘米蛋糕

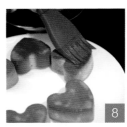

Two Hearts

Production workflow

1. Make the almond cake.
2. Make the raspberry butter cream.
3. Make the raspberry syrup.
4. Make the Chantilly cream.
5. Assemble the cakes

Almond Cake ❶

Ingredients

250 g marzipan ❷ (sweetened almond paste; 50% almond content)

20 g eggs

40 g egg yolk

70 g egg whites

5 g sugar

25 g cake flour

25 g cornflour

25 g Kirsch (cherry brandy)

60 g unsalted butter (melted and kept warm)

Method

1. Grease the moulds. Sieve flour and cornflour together.

2. In a mixing bowl, blend the eggs, egg yolk and marzipan with a bamix and beat until creamy and pale.

3. In another bowl, beat sugar and egg whites until stiff. Fold the egg whites gently into the egg yolk mixture.

4. Stir in flour and Kirsch. Add butter. Stir well.

5. Fill the moulds up to 80% of their depth. Smooth the top with a palette knife if needed. Cover the moulds with a non-stick baking lining or a silicone mat and then another baking tray so that the surfaces of the cakes stay flat as they rise. (see pictures 1-2)

6. Bake in a preheated oven at 180°C for 20 minutes until golden. Set aside to let cool. (see picture 3)

Raspberry Butter Cream

Ingredients

200 g raspberry puree

250 g sugar (A)

120 g eggs

30 g sugar (B)

300 g soft butter

1 vanilla pod (split open; seeds scraped out with a small knife)

Method

1. Boil the raspberry puree with sugar (A).
2. Mix the eggs and sugar (B) together. Pour in the boiling raspberry puree. Mix well. (see picture 4)
3. Heat the egg mixture up over a hot water bath while stirring constantly until it reaches 85°C. Pass the mixture through a sieve. (see picture 5)
4. Wait till the raspberry custard turns cold. Stir in the butter and beat well. Stir in the vanilla seeds at last. (see pictures 6-7)

Raspberry Syrup

Ingredients

100 g raspberry puree

60 g sugar

70 g water

Method

Boil all ingredients together. Leave it to cool.

Chantilly Cream

Ingredients

100 g whipping cream

20 g sugar

Method

Mix all ingredients and beat until stiff.

Assembly

1. Brush the cake with the raspberry syrup. (see picture 8)
2. Pipe raspberry butter cream and Chantilly cream over the cake. Garnish with fresh raspberries and edible gold leaf. (see pictures 9-10)

** Makes 9 cakes, measuring 5 cm x 6 cm x 3 cm each.

Tips

1 This almond cake has similar texture as a butter cake. After the egg whites are stirred into the batter, their bubbles do not raise the cake that much. So the cake will not be springy and fluffy like a sponge.

2 Raw marzipan is an almond confection made with ground almond and sugar. For this recipe, try to get marzipan with at least 50% almond content. The higher the almond content, the less sugar there is and the stronger the almond flavour will be. Some marzipan is specified to be used as decoration, with only 20 to 25% of almond content. Such marzipan is whiter in colour and extremely sweet so that it is not supposed to be used in a cake batter.

焦糖蘋果批
Caramel Apple Pie

焦糖與蘋果恆來相配；以焦糖慕絲、炒青蘋果加上合桃脆片，滿口清新又柔順的馨香。

炒青蘋果粒

Sautéed Diced Apple

材料

青蘋果 ❶ (切粒)	4 個
無鹽牛油	50 克
砂糖	50 克
蘋果酒	適量
檸檬汁	適量

做法

1. 用牛油和砂糖將青蘋果粒煮至焦糖化及收乾水。❷

2. 加入適量蘋果酒和檸檬汁,放涼待用。

焦糖慕絲

Bavarian Caramel Cream

材料

淡忌廉	70 克
牛奶	70 克
蛋黃	40 克
砂糖	48 克
魚膠片 (用凍水浸軟)	5 克
打起淡忌廉	125 克

做法

1. 牛奶和淡忌廉煮熱。

2. 用另一鍋將砂糖溶化,並煮至約 180℃ 使之變成焦糖。❸

3. 將 (2) 加入 (1) 內,然後倒入蛋黃內,以中快速攪打。

4. 過隔篩後加熱至 85℃,加入已浸軟的魚膠片,拌勻,待涼。輕
   輕拌入打起淡忌廉。 (看圖 1-2)

合桃脆片底 ❹
Walnut Brittle Base

材料

無鹽牛油	60 克
黃糖	40 克
葡萄糖膠	40 克
低筋麵粉	25 克
合桃 (烘香後放涼)	100 克

做法

1. 烘香合桃切成粉狀細末。

2. 牛油溶化，加入所有材料拌勻。

3. 將 (2) 放於兩張矽膠蓆或不黏布中擀平，再用 160℃焗約 12 分鐘。（看圖 3-4）

4. 焗好稍稍晾涼後，將脆片切出所需形狀。

蘋果球裝飾
Sautéed Apple Balls

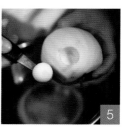

材料

青蘋果	4 個
無鹽牛油	40 克
砂糖	50 克

做法

用瓜剗將青蘋果挖成小球狀。用牛油和砂糖將它煮 至焦糖化，放涼待用。（看圖 5-6）

噴槍用白朱古力 ❺
White Chocolate Spray

材料

白朱古力	200 克
谷咕油	220 克

做法

將材料隔水或用微波爐慢慢加熱溶化至約 50℃，過濾，保持溫度。

裝組

1. 將焦糖慕絲注入模中至半滿，排上一層炒青蘋果粒，再舀焦糖慕絲至滿，放在冰格冷凍至結冰。（看圖 7-9）

2. 將 (1) 脫模，上面噴上一層白朱古力，放在合桃脆片上面。再伴以蘋果球及金箔作裝飾。（看圖 10）

** 可製 9 件 12 厘米 x 4 厘米 x 2.5 厘米的焦糖蘋果批

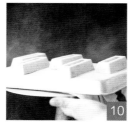

貼士

❶ 選用青蘋果是因為它不易煮軟，保持爽口，亦可選用金蘋果或日本玉林蘋果。

❷ 炒青蘋果前，煎鍋不需要預先燒熱，過熱只會燒焦牛油。待牛油稍溶化，就可下糖一起煮，煮至糖開始溶，就可下蘋果，勿待糖完全溶掉才下蘋果，因為糖漿接觸蘋果後會變硬塊，要花很多時間煮才能令變硬糖塊溶化，蘋果就會變得過軟。用大火把蘋果煮至收乾水就可以了。

❸ 煮焦糖要預備一隻厚底不銹鋼煲或銅煲，一杯水和一支毛筆。厚底不銹鋼煲或銅煲傳熱平均，糖會慢慢溶化、變焦；煮糖時不要攪伴，只需輕輕搖晃鋼煲，如看到有糖水在煲邊結成晶體，便用毛筆點少許水去擦掉，可防止晶體不溶，影響焦糖。

❹ 合桃脆片要在烘透後立即移到網架上放涼，否則釋出的油分會令脆片變軟，要再翻烘才會回復香脆。

❺ 如果沒有噴槍，可買市售即用朱古力裝飾噴劑或省去。

Caramel Apple Pie

Production workflow

1. Sauté the diced apple.
2. Make the Bavarian caramel cream.
3. Make the walnut brittle base.
4. Sauté the apple balls.
5. Make the white chocolate spray.
6. Assemble the pies. Spray white chocolate and garnish with apple balls.

Sautéed Diced Apple

Ingredients

4 granny smith apples ❶ (diced)
50 g unsalted butter
50 g sugar
Calvados (apple brandy)
lemon juice

Method

1. In a pan, cook the diced apples in butter and sugar until caramelized. ❷
2. Add Calvodos and lemon juice. Set aside and let cool.

Bavarian Caramel Cream

Ingredients

70 g whipping cream
70 g milk
40 g egg yolk
48 g sugar
5 g gelatine leaf (soak in cold water until soft, drained)
125 g whipped cream

Method

1. Heat the milk and the whipping cream until hot.
2. In another saucepan, melt the sugar until caramelized (about 180°C). ❸

3. Deglaze the saucepan with hot milk and cream. Mix well. Then pour the mixture into the egg yolk while beating vigorously.
4. Sift the mixture. Heat up to 85°C. Add the soaked gelatine leaf and stir until it dissolves. Leave it to cool. Fold in whipped cream. (see pictures 1-2)

Walnut Brittle Base ❹

Ingredients

60 g unsalted butter
40 g brown sugar
40 g glucose syrup
25 g cake flour
100 g walnuts (toasted and let cool)

Method

1. Chop the walnuts finely.
2. Melt butter and stir in all ingredients.
3. Spread the mixture between two silicone mats and bake in an oven at 160°C for about 12 minutes. (see pictures 3-4)
4. Cut into desired shapes whilst still warm.

Sautéed Apple Balls

Ingredients

4 granny smith apples
40 g unsalted butter
50 g sugar

Method

Scoop out apple balls with a melon baller. In a pan, cook them with butter and sugar until caramelized. Set aside. (see pictures 5-6)

White Chocolate Spray ❺

Ingredients

200 g white chocolate

220 g cocoa butter

Method

Melt all ingredients over a hot water bath or in the microwave to about 50°C. Pass it through a sieve. Keep the temperature.

Assembly

1. Spoon the Bavarian caramel cream into the ingot mould until half full. Arrange a layer of sautéed diced apple. Top with remaining Bavarian caramel cream until full. Freeze it. (see pictures 7-9)

2. Unmould the caramel cream and spray white chocolate on top with a chocolate sprayer. Place it on a piece of walnut brittle base. Garnish with sautéed apples balls and edible gold leaf. (see picture 10)

** Make 9 rectangular mini pies, measuring 12 cm x 4 cm x 2.5 cm each.

Tips

1. Granny smith apples are used in this recipe because they stay crunchy and firm even after prolonged cooking. You may also use golden delicious or Japanese Ourin apples.

2. When you sauté the apples, do not heat up the pan first. A pan too hot might burn the butter easily. Heat until the butter melts. Add sugar and cook until the sugar begins to melt. Put in the apples at this point. Do not wait until the sugar is melted because the molten sugar is considerably hotter than the apples. A layer of sugar will solidify on the surface of the apples. It will take much time to cook through the apples this way. Just cook the apples over high heat until the syrup reduces.

3. To make caramel, you need a heavy-bottom pot or a copper pot, a glass of water and a brush. A heavy-bottom or copper pot conducts heat evenly so that the sugar will melt and caramelize slowly. Do not stir the syrup when you cook it. Just swirl the pot slowly. In case the syrup starts to crystallize on the rim of the pot, dip a brush in water and brush the crystals back in the syrup. This would prevent the syrup from crystallizing which might make the caramel lumpy.

4. After baking the walnut brittle base, transfer immediate onto a cooling rack. Otherwise, the oil from the walnut brittle would make itself soggy. It needs to be re-baked to be crispy again.

5. If you don't have a chocolate sprayer, get readymade chocolate spray in a can. Or simply skip that part.

鳥結香橙條
Nougat Orange Slice

杏仁可説是法國甜品的靈魂，餅乾、燒果子，無處不聞其
馨香；香味含蓄又細膩的鳥結醬更是法國甜品的繆思。

製作次序

1. 橙醬、橙味餡心
2. 杏仁鬆酥
3. 鳥結醬慕絲
4. 裝組

橙醬

Orange Marmalade

材料

橙	250 克
檸檬汁	1/2 個
橙汁	250 克
砂糖	150 克

做法

1. 切去橙蒂，將橙連皮切成大塊，去籽，放在鍋內，加水至剛浸過所有材料，煮約 15 分鐘除去苦味，用清水沖洗。

2. 橙汁、糖、橙塊一同放進鍋內煮約一小時，期間用隔油湯杓濾掉細沫，煮至液體有光澤及濃稠，滴在碟子上不會化開。

3. 加入檸檬汁調味。

橙味餡心

Orange Jelly

材料

橙醬	200 克
魚膠片（用凍水浸軟）	3 克
橙酒	5 克
水	30 克

** 可用 150 克 Boiron 半糖漬香橙代替橙醬，但水要再加添多 50 克。

做法

將水煮滾，加入已浸軟魚膠片及其餘材料，拌勻，然後注入容器內，放入冰格冷凍至結冰。
（看圖 1-2）

杏仁鬆酥
Almond Shortbread Pastry

材料

無鹽牛油（室溫放軟）	50 克
糖霜	30 克
雞蛋	20 克
低筋麵粉	87 克
杏仁粉	12 克
發粉	1 克
鹽	1 克
烘香杏仁片	26 克
溶化白朱古力	適量

做法

1. 牛油、糖霜及鹽用膠刮輕輕拌勻至糖霜被牛油吸收。

2. 逐少加入雞蛋，攪拌至蛋漿被牛油糖糊完全吸收。

3. 加入已篩勻的粉類和烘香杏仁片，拌勻至粉糰狀。

4. 將粉糰放入雪櫃，冷凍至結實，取出；用擀麵棒將粉糰打至略平，再擀薄至厚 2 毫米的餅皮，再切出和模具相約的形狀。（看圖 3-4）

5. 放入已預熱 180℃ 的焗爐內，焗約 15-18 分鐘至金黃色。（看圖 5）

6. 待涼後，塗上溶化白朱古力，冷藏待用。

鳥結醬慕絲
Nougat Cream

材料

牛奶	130 克	吉士粉	20 克
淡忌廉	120 克	魚膠片 (用凍水浸至軟)	4 克
蛋黃	80 克	鳥結醬	200 克
砂糖	80 克	打起淡忌廉	250 克

做法

1. 牛奶和淡忌廉一起煮滾，關火。

2. 將蛋黃、吉士粉和糖拌勻，將 (1) 逐少倒入蛋糖料內（期間要不斷攪拌），然後蛋奶混合料倒回鍋內，用慢火煮至變稠（期間要不斷攪拌），離火，過濾。（看圖 6-7）

3. 趁熱加入鳥結醬和已浸軟的魚膠片，待涼後拌入打起淡忌廉。（看圖 8-9）

裝組

1. 先放一層鳥結醬慕絲進模內，放入餡心，再填滿鳥結醬慕絲，放入冰格冷凍至結冰。（看圖 10-12）❷

2. 鳥結醬慕絲脫模，然後噴上朱古力裝飾，再放在杏仁脆餅上即成。

* 朱古力噴飾，參考第 30 頁的栗子忌廉黃豆粉蛋白餅。

貼士

❶ 鳥結醬是以杏仁加糖炒香後，碾磨而成的果仁醬。

❷ 如用矽膠模，可直接將慕絲入模；但如用金屬餅圈，則需要用保鮮膜包實圈底以防滲漏。

Nougat Orange Slice

Production workflow

1. Make the orange marmalade and then the orange jelly.
2. Make the almond shortbread pastry.
3. Make the nougat cream.
4. Assemble and spray on chocolate as garnish.

Orange Marmalade

Ingredients

250 g orange
juice of 1/2 lemon
250 g orange juice
150 g sugar

Method

1. Remove the stem of the orange. Cut into large chunks with skin on. Seed them. Place in a pot. Add water to cover the orange chunks. Boil for 15 minutes to remove its bitterness. Rinse in water. Drain.
2. In a pot, put in orange juice, sugar and orange chunks. Cook for 1 hour. Skim off the foam on the surface with a fine mesh strainer ladle throughout the cooking process. Cook until the liquid is shiny and thick. When dripped on the counter, it should form a drop without running flat.
3. Season with lemon juice.

Orange Jelly

Ingredients

200 g orange marmalade*

3 g gelatine leaf (soak in cold water until soft; drained)

5 g Cointreau (orange liqueur)

30 g water

*To save time and effort, you may use 150 g semi-candied orange of Boiron brand instead of 200 g of orange marmalade. But make sure you add 50 g more water.

Method

Bring water to the boil. Add gelatine leaf, orange marmalade and Cointreau. Pour into the mould and keep in freezer until frozen. (see pictures 1-2)

Almond Shortbread Pastry

Ingredients

50 g unsalted butter (at room temperature)
30 g icing sugar
20 g egg
87 g cake flour
12 g ground almonds
1 g baking powder
1 g salt
26 g toasted flaked almonds
molten white chocolate

Method

1. Mix butter, icing sugar and salt together with a plastic spatula until the icing sugar is moistened by the butter.
2. Add the egg little by little and stir well after each addition. Stir until well incorporated.
3. Put in the dry ingredients and toasted almonds. Stir into dough.
4. Refrigerate the dough until firm. Bash it gently with a rolling pin to flatten slightly. Roll it out with a rolling pin into a dough sheet about 0.2 cm thick. Cut it into pieces according to the size of the mousse ring you use. (see pictures 3-4)
5. Bake in a preheated oven at 180°C for 15 to 18 minutes until golden. (see picture 5)
6. Leave it to cool. Brush the molten white chocolate over the cooled pastry. Refrigerate for later use.

Nougat Cream

Ingredients

130 g milk

120 g whipping cream

80 g egg yolk

80 g sugar

20 g custard powder

4 g gelatine leaf (soaked in water until soft; drained)

200 g nougat paste❶

250 g whipped cream

Method

1. In a pot, put in the milk and whipping cream. Bring to the boil. Turn off the heat.

2. In a mixing bowl, whisk the egg yolk, custard powder and sugar together. Pour the hot cream mixture from step 1 into this egg yolk mixture while stirring continuously. Pour the resulting mixture back into the pot. Put it over low heat while stirring constantly until it boils. Remove from heat and pass it through a sieve. (see pictures 6-7)

3. Stir in the nougat paste and gelatine leaf. Stir until gelatine dissolves. Leave it to cool completely before folding in whipped cream. (see pictures 8-9)

Assembly

1. In a silicone mould or mousse ring, spread a layer of nougat cream on the bottom first. Put the frozen orange jelly over the mousse. Fill the mould or ring up with the remaining mousse. Keep in the freezer until frozen. (see pictures 10-12) ❷

2. Unmould and spray on a layer of chocolate with a chocolate sprayer (or chocolate spray in a can). Transfer onto a piece of almond shortbread. Serve.

* Refer to Chestnut Cream with Soya Dacquoise on p.30 to prepare the chocolate for sprayer.

Tips

❶ Nougat paste is made from almonds that are toasted with sugar and then ground into a paste.

❷ You have to wrap the mousse ring well to prevent the mousse from leaking. If you use a silicone mould, you may skip this step.

黑加侖子芝士蛋糕
Blackcurrant Cheese Cake

很簡單的食譜，來自加拿大的大舅母。無論弄甚麼芝士餅給兒子吃，他總是
說這個最好吃。我配了酸酸甜甜的黑加侖子醬，令這芝士餅更清爽對胃。

製作次序	1. 黑加侖子醬	3. 芝士糊
	2. 餅底	4. 裝組

黑加侖子醬
Blackcurrant Sauce

材料

黑加侖子果茸或果汁	250 克
砂糖	20 克
粟粉	10 克

做法

將所有材料混合，用小煲以細火煮滾。倒
入不黏模或圓形模具（高約 0.5 厘米，寬
約 3 厘米）中冷凍至結冰。

餅底
Crust

材料

消化餅	100 克
牛油 (室溫放軟)	約 50 克

做法

1. 消化餅壓碎，與軟牛油拌勻。

2. 在模具上塗上軟牛油（份量外）[1]，撒
   上一層薄筋粉（份量外），在模底墊
   一層牛油紙，把餅碎用湯匙壓在餅模
   底。

芝士糊
Cheese Filling

材料

忌廉芝士	500 克
雞蛋	180 克
砂糖	90 克
天然雲呢拿香油	數滴

做法

1. 用食物處理器將所有材料拌勻。 ❷

2. 芝士糊倒入餅模內，放進已預熱 130℃ 的焗爐，焗約 30 分鐘，待涼。

裝組

在雪凍芝士餅上放黑加侖子醬和黑莓裝飾。

** 可做約 10 個 5.5 厘米寬 3.5 厘米高的小餅

貼士

❶ 把模具塗上軟牛油，再撒上一層薄筋粉，可令芝士餅容易出模和周邊滑溜。

❷ 芝士糊勿在食物處理器內攪打太久，一但過熱，芝士糊便會變稀，影響成品。沒有食物處理器的讀者，可用電動手提攪拌器或電動棒狀攪拌器。

Blackcurrant Cheese Cake

Production workflow

1. Make the blackcurrant sauce.
2. Make the crust.
3. Make the cheese filling.
4. Assemble and serve.

Blackcurrant Sauce

Ingredients

250 g blackcurrant puree or blackcurrant juice

20 g sugar

10 g cornflour

Method

Mix all ingredients together and put into a small pot. Heat over low heat until it boils. Divide among ten 3 cm wide round moulds (about 0.5 cm thick) or pour on non-stick moulds. Keep in freezer until frozen.

Crust

Ingredients

100 g digestive biscuit

about 50 g butter (at room temperature)

Method

1. Crush the biscuits. Mix well with butter until crumbly.
2. Grease ten 5.5 cm x 3.5 cm round moulds with butter and dust them with cake flour. ❶ Then line them with greaseproof paper on the bottom. Press the biscuit mixture with a tablespoon onto the bottom of the moulds.

Cheese Filling

Ingredients

500 g cream cheese

180 g eggs

90 g sugar

a few drops natural vanilla extract

Method

1. In a food processor, blend all ingredients together. ❷
2. Pour into the moulds over the crust. Bake in a preheated oven at 130°C for about 30 minutes. Leave them to cool.

Assembly

Refrigerate the cheese cakes until set. Place the frozen blackcurrant sauce on top. Garnish with a blackberry. Serve.

** Makes 10 mini cheese cakes, measuring 5.5 cm x 3.5 cm each.

Tips

❶ Greasing the moulds with butter and dusting them thinly with cake flour helps unmould more easily. The cheese cakes tend to have smooth edges too.

❷ Do not over-beat the cheese filling in the food processor. Heat builds up in the mixing process. The cream cheese will go watery once over-heated, so that the cheese cakes will not be smooth. Those readers who don't have a food processor may use an electric mixer or a hand blender.

海鹽焦糖榛子薩瓦林鬆酥

*Salted Caramel
Hazelnut Gourmandise*

Tenderness 柔軟的內涵深受網友喜愛，使薩瓦林 (Savarin) 模具風行一時。薩瓦林確是很聰明的設計，稍花心思即可變化萬千。這食譜再次巧用薩瓦林的特點，把榛子與焦糖兩種天衣無縫的食材結合。

製作次序		
1.	榛子鬆酥	4. 裝組
2.	海鹽焦糖拖肥	
3.	焦糖榛子	

榛子鬆酥

Hazelnut Vanilla Shortbread

材料

無鹽牛油（室溫放軟）	150 克
雲呢拿豆莢	1 枝（用小刀刮出雲呢拿籽）
轉化糖	5 克
糖霜	60 克
蛋黃	50 克
鹽	2 克
低筋麵粉	125 克
發粉	5 克
榛子粉	40 克

做法

1. 用膠刮將牛油、糖霜、轉化糖及雲呢拿籽混合成軟糊。

2. 加入蛋黃和鹽，拌勻。（看圖 1）

3. 低筋麵粉及發粉過篩，然後拌入牛油蛋糊內。注意切勿攪拌過度。（看圖 2）

4. 加入榛子粉，拌勻。

5. 將麵糊舀入模中至八成滿，把模具摔在桌上數下讓麵糊緊貼模具。蓋上不黏布或矽膠蓆，再壓上焗盤及重物，令蛋糕發起時底部平整。（看圖 3-7）

6. 放入已預熱 180℃ 的焗爐內，焗約 20 分鐘。取出待涼，脫模。（看圖 8）

海鹽焦糖拖肥
Salted Butter Caramel

材料

砂糖	75 克
葡萄糖漿	60 克
蜂蜜	12 克
淡忌廉（煮熱）	100 克
無鹽牛油（室溫放軟）	20 克
雲呢拿豆莢	1/2 枝（用小刀刮出雲呢拿籽）
海鹽	2 克

做法

1. 砂糖、葡萄糖及蜂蜜倒入厚底的不鏽鋼煲煮至微黃。

2. 離火，加入熱忌廉、牛油、海鹽及雲呢拿籽，繼續煮至 114℃。離火，待用，須保溫。（看圖 9-10）

焦糖榛子
Caramelized Hazelnuts

材料

益壽糖	約 200 克
烘香榛子	約 10 粒

做法

1. 益壽糖放在厚底不銹鋼煲內煮至微焦，放涼至糖膠變得較濃稠。（看圖 11）

2. 用長竹簽戳住榛子，蘸滿糖膠，將長竹簽掛在枱邊或高處，任由糖膠流下。（看圖 12）

3. 待糖膠變硬，用剪刀修剪成所需長度。（看圖 13）

裝組

1. 榛子鬆酥涼透後，用粉篩磨去突出、多餘的邊緣。（看圖 14-15）

2. 將海鹽焦糖拖肥舀進榛子鬆酥內，放上焦糖榛子裝飾。（看圖 16-17）

** 可做 9 個 8 厘米 x 6 厘米 x 2.5 厘米薩瓦林蛋糕

貼士

① 不要將焦糖煮至太深色，否則焦糖會很苦。

② 要逐少逐少加入熱忌廉，大量地加入會令焦糖過熱而令忌廉沸騰瀉出，構成危險。

Salted Caramel Hazelnut Gourmandise

Production workflow

1. Make the hazelnut vanilla shortbread.
2. Make the salted butter caramel.
3. Make the caramelized hazelnuts.
4. Assemble.

Hazelnut Vanilla Shortbread

Ingredients

150 g unsalted butter (at room temperature)

1 vanilla pod (split open; seeds scraped out with a knife)

5 g invert sugar

60 g icing sugar

50 g egg yolk

2 g salt

125 g cake flour

5 g baking powder

40 g ground hazelnuts

Method

1. Mix butter with icing sugar, invert sugar and vanilla seeds with a plastic spatula.
2. Add egg yolk and salt. Mix well.
3. Sift in cake flour and baking powder together. Fold into the egg mixture from step 2. Do not over-stir.
4. Add ground hazelnuts. Mix well.
5. Fill the Savarin moulds with the batter up to 80% of their depth. Tap the mould against the counter gently for a few times to release any trapped bubble. Cover the moulds with a non-stick baking lining or a silicone mat. Place something heavy on top so that the base of the Savarin cakes will be flat when they rise.
6. Bake in a preheated oven at 180°C for about 20 minutes. Let them to cool and unmould.

**(see pictures 1-8)

Salted Butter Caramel

Ingredients

75 g sugar

60 g glucose

12 g honey

100 g whipping cream (heated through)

20 g unsalted butter

1/2 vanilla pod (split open; seeds scraped out with a knife)

2 g sea salt

Method

1. In a heavy-bottom saucepan, cook sugar, glucose and honey until light brown. ❶
2. Remove from heat. Pour in hot cream, ❷ butter, sea salt and vanilla seeds. Continue cooking until the caramel reaches 114°C. Set aside and keep warm. (see pictures 9-10)

Caramelized Hazelnuts

Ingredients

200 g isomalt

10 toasted hazelnuts

Method

1. Cook isomalt in a heavy-bottom saucepan until lightly browned. Leave it to cool until it thickens.
2. Secure one hazelnut on the tip of each bamboo skewer. Dip the hazelnuts into the molten syrup. Hang the bamboo skewer on the edge of the counter and let the excess syrup drip down.
3. Wait until the syrup cools and hardens. Trim with a pair of scissors into desired length.

**(see pictures 11-13)

Assembly

1. Wait until the shortbread is cool. Use a sieve or a grater to file off the rugged and irregular edge. (see pictures 14-15)

2. Pour some salted butter caramel into the notch of the shortbread. Garnish with a caramelized hazelnut. (see pictures 16-17)

** Makes 9 Savarin cakes, measuring 8 cm x 6 cm x 2.5 cm each.

Tips

1 Cook the caramel to light brown. It tastes bitter if it is cooked to a darker colour.

2 Be careful when you pour in the hot cream into the caramel. The caramel is very hot and the cream may overflow and spill.

乳酪乾果蛋糕
Yoghurt Dried Fruit Oeuf

燒果子的魅力在於濃郁，對有些人來說也失之於厚重——重牛油、重糖、重麵粉，紮實的口感也讓人有很大壓力。此食譜以大量酸乳酪為基底，加上清香的冧酒糖水漬果乾、椰茸，清爽輕盈——先旨聲明，是口味，不是卡路里。

製作次序
1. 冧酒糖水漬果乾
2. 酸乳酪風味蛋糕
3. 裝組

冧酒糖水漬果乾
Rum-infused Dried Fruits

材料

提子乾	50 克
杏脯	120 克
冧酒	20 克
砂糖	30 克
水	100 克

做法

所有材料一同煮約 15-20 分鐘，放涼，切碎。（看圖 1-2）

酸乳酪風味蛋糕

Yoghurt cake

材料

冧酒糖水漬果乾	全部	高筋麵粉	90 克
無鹽牛油	100 克	低筋麵粉	130 克
砂糖	140 克	椰絲	70 克
雞蛋	120 克	海鹽	1 克
原味乳酪	250 克	天然雲呢拿香油	
發粉	3 克	杏桃 (裝飾用)	
梳打粉	2 克	紅加侖子 (裝飾用)	

做法

1. 砂糖與牛油打發至鬆發變白,然後加入蛋及乳酪拌勻。

2. 輕輕拌入其餘的乾性材料及雲呢拿香油。注意不要攪拌過度。

3. 最後加入冧酒糖水漬果乾,稍為拌勻。(看圖 3)

4. 將麵糊舀入模中至八成滿,需要時用小拉刀抹平。蓋上不黏布或矽膠蓆,壓上焗盤,令蛋糕底部平滑。(看圖 4-5)

5. 放入已預熱 180℃ 的焗爐內焗約 20 分鐘至金黃色。(看圖 6)

裝組

杏桃抹乾水分,蘸上砂糖,放在焗盤上,以火槍燒至金黃色,然後放在酸乳酪風味蛋糕上,再加上紅加侖子。

Yoghurt Dried Fruit Oeuf

Production workflow

1. Steep the dried fruits in rum syrup.
2. Make the yoghurt cake.
3. Assemble the cakes.

Rum-infused Dried Fruits

Ingredients

50 g raisins

120 g dried apricots

20 g rum

30 g sugar

100 g water

Method

Cook all ingredients in a pot for 15 to 20 minutes. Leave it to cool. Chop finely. (see pictures 1-2)

Yoghurt Cake

Ingredients

rum-infused dried fruits

100 g unsalted butter

140 g sugar

120 g eggs

250 g plain yoghurt

3 g baking powder

2 g baking soda

90 g bread flour

130 g cake flour

70 g grated coconut

1 g sea salt

vanilla essence

apricots (as garnish)

red currants (as garnish)

Method

1. Beat sugar and butter until light and fluffy. Stir in eggs and plain yoghurt.
2. Then fold in all dry ingredients and vanilla essence. Be careful not to over-stir the batter.
3. Stir in the dried fruits gently. (see picture 3)
4. Fill the oval moulds up to 80% of their depth. Smooth the top with a palette knife if needed. Cover the moulds with a non-stick baking lining or a silicone mat. Put a baking tray on top, so that the bases of the cakes stay flat as they rise. (see pictures 4-5)
5. Bake in a preheated oven at 180°C for about 20 minutes until golden brown. (see picture 6)

Assembly

Wipe dry the apricots. Dip them in sugar. Put them on a baking tray. Caramelize with a propane torch. Arrange on the oval yoghurt cakes. Then garnish with red currants.

心太軟
Chocolate Moelleux

「心太軟」這名字既形似，也具裊裊的弦外之意，絕妙。

這鬆餅般的暖朱古力布丁，要外脆內軟，內裏卻像天鵝絨柔軟、醇香、濕潤，叉子戳進去時有如火山溶岩般流出的震撼。佐冰凍的雪糕同吃，冷、熱、脆、軟四種口感衝擊著味蕾，方算淋漓盡致。

製作次序	1. 朱古力餡心
	2. 朱古力布丁糊
	3. 裝組

朱古力餡心
Ganache Filling

材料

淡忌廉	30 克
黑朱古力 (50% 谷咕含量以上) ❶	20 克
牛油 (室溫放軟)	5 克

做法

1. 淡忌廉加熱，倒入朱古力內，融化後拌勻。

2. 加入軟牛油，拌勻，注入小圓矽膠模內，放入冰格雪至結冰待用。(附圖)

朱古力布丁糊 ❷
Chocolate Moelleux

材料

黑朱古力 (50% 谷咕含量以上)	140 克
無鹽牛油	120 克
雞蛋	140 克
砂糖	80 克
低筋麵粉 (篩勻)	60 克

做法

1. 牛油和朱古力一同隔水或用微波爐加熱溶化，涼至室溫。

2. 雞蛋及砂糖打起至濃稠變淺色。

3. 輕輕把已篩麵粉拌入 (2)，再拌入朱古力牛油混合物。將麵糊雪藏最少一小時甚至一晚。

裝組

1. 金屬模具包上錫紙,塗油,墊上牛油紙。

2. 用雪糕杓把麵糊舀進模具至半滿,釀入已冷藏結冰的朱古力餡心,再填滿麵糊。

3. 用 180℃ 焗約 13-14 分鐘,出爐。❸ 配以雪糕一起享用。

** 看圖 1-7

** 可做 6 個約 100 克重的布丁

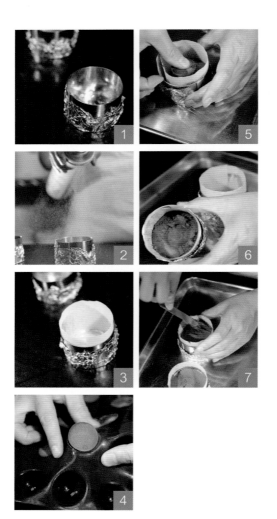

貼士

❶ 食譜中,朱古力後面括弧內的是所含谷咕量的比率,大多數的優質朱古力都有說明所含谷咕量的比率,含谷咕量越高糖分則越少。

❷ 朱古力布丁糊做好後最少要雪上一小時或過夜才使用,焗的時候才不會發得過高。

❸ 這布丁製作的程序不是最重要,最重要的是掌握好烤焗的時間,要熟悉自己焗爐的火力。全個布丁約焗 13-14 分鐘,目的是把布丁糊焗至七‧八成熟,而釀在裏面的餡心剛遇熱溶解就可以了。烘焙過久的話,餡心會被布丁完全吸收;時間過短則布丁未熟,未能成形,布丁會因過軟而崩塌。

Chocolate Moelleux

Production workflow

1. Make the ganache filling.
2. Make the chocolate moelleux.
3. Assemble. Bake. Serve.

Ganache Filling

Ingredients

30 g whipping cream
20 g dark chocolate (over 50% cocoa)❶
5 g butter (at room temperature)

Method

1. Heat the cream then pour over the chocolate. When chocolate melts, mix well.
2. Add butter and mix well. Freeze it. (see picture in p.81)

Chocolate Moelleux❷

Ingredients

140 g dark chocolate (over 50% cocoa)
120 g unsalted butter
140 g eggs
100 g sugar
60 g cake flour (sifted)

Method

1. Melt the chocolate with butter over a hot water bath or in a microwave. Leave it to cool at room temperature.
2. Whisk the eggs with sugar until light and fluffy.
3. Fold flour into the egg mixture. Then stir in chocolate mixture. Mix well and refrigerate for at least 1 hour or overnight.

Assembly

1. Line metal rings with aluminium foil. Grease them and line with parchment paper.
2. Use an ice-cream scoop to fill the rings with chocolate cake batter up to half-full. Put in the frozen ganache filling. Top with more chocolate cake batter until full.
3. Bake in a preheated oven at 180°C for 13 to 14 minutes.❸ Turn them out onto a serving plate and serve with a scoop of ice cream.

** see picture 1-7
** Makes 6 cakes, about 100 g each.

Tips

❶ Whenever chocolate is called for in a recipe, it is followed by a specified percentage of cocoa. Most quality cooking chocolate has cocoa content listed on its package. The higher the percentage of cocoa is in a chocolate, the less sugar it contains.

❷ After assembling the cake, refrigerate for at least one hour or overnight before baking. Otherwise, the batter might rise too high when baked.

❸ The key to this recipe lies in the baking step. Control the baking time well and understand the power of your oven. The perfect moelleux au chocolat should have a cooked crust and a gooey runny centre. Thus, the cake batter should be baked so that it is about medium-to-well done on the outside and the ganache is just hot enough to melt. Over-baking means the ganache runs and will be picked up completely by the cake. Under-baking means a firm crust of cake has not been formed yet and the cake may not be able to stand in one piece when turned out from the ramekin.

南瓜吉士卷
Pumpkin Custard Rouge

番薯、香芋等富含澱粉質的根菜，向來教女士喜愛；南瓜豐盈而細緻的質感、獨特、富泥土氣息的甘甜，讓它成為甜點中獨領風騷的蔬菜。通常用來配合燒果子，今次做成輕盈的卷蛋，味與形俱妙。

南瓜卷蛋
Pumpkin Sponge Cake Roll

材料

蛋黃	140 克
砂糖 (A)	12 克
蛋白	150 克
砂糖 (B)	50 克
低筋麵粉	35 克
蒸熟日本南瓜茸 ❶	30 克
無鹽牛油（溶化）	15 克
牛奶	20 克
日本南瓜（切薄片）❷	1/4 個
日本南瓜籽	適量

做法 ❸

1. 南瓜切成薄片，修剪成彎月形，排放在膠蓆上，並撒上南瓜籽。

2. 牛油與牛奶一同煮熱。

3. 蛋黃與砂糖 (A) 打發至濃稠。

4. 蛋白與砂糖 (B) 打至企身。

5. 將蛋黃糊拌入蛋白糊，輕輕拌勻。

6. 輕輕拌入已篩麵粉。

7. 把小部份（6）加入南瓜茸及（2）內拌勻，然後倒回餘下的（6），徹底調勻。

8. 將 (7) 倒在膠蓆上，抹平，放入已預熱 180℃ 的焗爐內，焗約 15 分鐘。

** （看圖 1-10）

** 可製 30 厘米 x 40 厘米 x 1 厘米蛋糕 1 盤

日本南瓜吉士忌廉

Pumpkin Custard

材料

牛奶	100 克
雲呢拿豆莢	半支 （用小刀刮出雲呢拿籽）
蛋黃	20 克
砂糖	20 克
吉士粉	8 克
無鹽牛油（室溫放軟）	20 克
蒸熟日本南瓜茸	70 克
打起淡忌廉	50 克

做法

1. 牛奶和雲呢拿籽一起煮熱。

2. 蛋黃、砂糖及吉士粉混合，一邊攪拌一邊慢慢加入（1），拌勻，過篩。

3. 將（2）倒回鍋內加熱，煮滾至濃稠，關火，放涼至約 40℃，加入牛油和南瓜茸。

4. 待完全涼透後，拌入打起淡忌廉。

裝組 ❹

1. 蛋糕下墊一張牛油紙，噴上酒糖水增加濕潤。（看圖 11）

2. 塗上一層薄南瓜吉士忌廉，用一枝幼木棍協助把蛋糕捲好。（看圖 12-14）

3. 放進雪櫃冷藏定型。

貼士

❶ 南瓜預先切大件蒸熟，然後去皮壓成茸，放涼備用。宜選用日本出產的南瓜，它味道香濃粉甜，不用加任何食材已經很美味，就連瓜皮也可食用，勿浪費啊！

❷ 鋪在烤盤上的南瓜要切得薄一點，才容易烘熟。

❸ 分蛋法的海綿蛋糕體，要注意使用室溫雞蛋，並把牛奶和牛油加熱，麵糊才不會因牛油太冷而凝聚在麵糊底部。

❹ 卷蛋的餡料不能太厚，否則會擠出來，難以造型。可在開始捲的一端塗多些，在尾部塗薄一點。

Pumpkin Custard Rouge

Production workflow

1. Make the pumpkin sponge cake roll.
2. Make the pumpkin custard
3. Assemble

Pumpkin Sponge Cake Roll

Ingredients

140 g egg yolks
12 g sugar (A)
150 g egg whites
50 g sugar (B)
35 g cake flour
30 g steamed mashed Japanese pumpkin ❶
15 g unsalted butter (melt)
20 g milk
1/4 Japanese pumpkin (sliced) ❷
pumpkin seeds

Method ❸

1. Slice the pumpkin thinly. Trim into a crescent shape. Arrange on a silpat lined baking tray. Sprinkle pumpkin seeds on top.
2. In a pot, heat the butter and the milk.
3. In a bowl, beat egg yolks and sugar (A) until pale and thick.
4. In another bowl, beat egg whites and sugar (B) until stiff.
5. Fold the egg yolk mixture from step 3 into the egg whites from step 4.
6. Sieve in the cake flour and fold well.
7. Put a small volume of the batter from step 6 into the hot milk mixture from step 2. Add mashed pumpkin and stir well. Then mix well with the remaining butter.
8. Pour the resulting batter into the silpat lined baking tray over the pumpkin slices and seeds. Bake in a preheated oven at 180°C for 15 minutes.

** (see pictures 1-10)
** Makes 30 cm x 40 cm x 1 cm sponge cake.

Pumpkin Custard

Ingredients

100 g milk

1/2 vanilla pod (split open; seeds scraped out with a knife)

20 g egg yolk

20 g sugar

8 g custard powder

20 g unsalted butter (at room temperature)

70 g steamed mashed Japanese pumpkin

50 g whipped cream

Method

1. In a pot, cook milk and vanilla seeds until it boils.

2. In a bowl, mix together egg yolk, sugar and custard powder. Pour in the hot milk mixture from step 1 while stirring continuously. Pass the mixture through a sieve.

3. Pour back into the pot and heat it until it boils. Stir constantly all along. Remove from heat and leave it to cool to 40°C. Stir in butter and mashed pumpkin.

4. Wait until it has cooled completely. Fold in the whipped cream.

Assembly ❹

1. Lay a sheet of baking paper on the counter first. Put the pumpkin sponge cake over it. Spray sugar syrup all over to keep it moist. (see picture 11)

2. Spread a layer of pumpkin custard over the sponge cake. Roll the cake up with the help of a small rolling pin. (see pictures 12-14)

3. Refrigerate until set.

Tips

❶ To make the mashed pumpkin, cut a pumpkin into large chunks first. Steam until soft. Peel and mash finely. Leave it to cool before use. I prefer Japanese pumpkin for its sweetness and abundant starch. It actually tastes great as it is. Even the skin is edible. Don't chute it.

❷ Those pumpkin slices placed on the baking tray should be very thinly sliced. Otherwise, they might not get cooked properly.

❸ The sponge cake is made with egg whites and yolks beaten separately. Make sure all eggs are brought to room temperature before use. The butter has to be cooked with the milk so that it won't separate and sink to the bottom of the batter.

❹ Do not spread too much custard on the sponge cake. Otherwise, the filling tends to be squeezed out as you roll it and mess up the presentation. You may spread more custard on the end that will be rolled in first. But apply less toward the other end.

檸檬椰子條
Lemon Coconut Slice

以市售半糖漬檸檬和檸檬醬製作，烤焗時檸檬醬滲進杏仁味香濃的鬆軟蛋糕內，加上稍有焦香又不失清甜的椰子茸，感覺清新輕爽。

製作次序
1. 半糖漬檸檬蛋糕
2. 椰子茸飾面

半糖漬檸檬蛋糕
Lemon Cake

材料

杏仁粉	95 克
糖霜	55 克
雞蛋	95 克
無鹽牛油 (溶化)	50 克
低筋麵粉	15 克
半糖漬檸檬或檸檬芳頓糖 ❶	20 克
檸檬醬	約 50 克

椰子茸飾面
Coconut Topping

材料

椰子茸	40 克
砂糖	10 克
雞蛋	25 克

做法

1. 將所有椰子茸飾面材料拌勻。

2. 杏仁粉、糖霜和雞蛋用高速攪拌，直至變白及鬆發，約 3 分鐘。

3. 轉低速，加入已過篩的麵粉及溶化牛油。注意不要打發過度。

4. 加入半糖漬檸檬或檸檬芳頓糖，拌勻。

5. 將麵糊舀進擠花袋內，然後將麵糊擠進矽膠模內，放入已預熱 180℃ 的焗爐內，焗約 10 分鐘。

6. 取出蛋糕，擠上檸檬醬，再放上椰茸飾面，放回焗爐繼續焗約 10 分鐘至金黃色。（看圖 1-3）

** 可製約 12 條檸檬椰子條

Lemon Coconut Slice

Lemon Cake

Ingredients

95 g ground almonds
55 g icing sugar
95 g eggs
50 g melted unsalted butter
15 g cake flour
20 g semi-candied lemon **1** or lemon fondant
50 g lemon curd

Coconut Topping

Ingredients

40 g desiccated coconut
10 g sugar
25 g egg

Method

1. Mix together all ingredients of the coconut topping.

2. In another mixing bowl, blend ground almonds, icing sugar and eggs over high speed for about 3 minutes until pale and light.

3. Turn to low speed and sieve in the flour. Add melted butter. Do not over-stir it.

4. Add semi-candied lemon or lemon fondant. Mix well.

5. Put the batter into a piping bag. Pipe it into the silicone moulds. Bake in a preheated oven at 180°C for 10 minutes.

6. Pipe some lemon curd on the cakes. Spread the coconut topping on top. Continue baking for 10 more minutes until golden. (see picture 1-3)

Tips

1 I recommend semi-candied lemon of the French brand Boiron for this recipe. If you can't get it, replace it with one part fresh lemon (skin on); one part icing sugar and one part fondant. Blend into a smooth paste consistency. It lasts in the fridge for a month.

** Makes about 12 rectangular cake strips.

貼士

1 宜使用法國 Boiron 品牌的半糖漬檸檬,如買不到,可以 1:1:1 份量的連皮鮮檸檬、糖霜和芳頓糖霜,用食物處理器攪成幼滑糊狀代替使用,可存放雪櫃約一個月。

谷咕果仁提子曲奇
Nibby Nut and Raisin Cookies

可可豆是一顆顆靜躺於可可樹果實內的小豆子，經過多番加工、去殼，製成朱古力、可可粉。後來有製造商發現跟可可豆胚一同經過發酵、烘焙的可可豆殼，氣味濃郁，甘脆可口，於是把它加工成另一種美味食材。

材料

無鹽牛油	115 克
砂糖	58 克
黃糖	58 克
海鹽	2 克
蛋	50 克
中筋麵粉	160 克
梳打粉	2 克
谷咕果殼碎	38 克
烘香合桃	58 克
提子乾	50 克

做法

1. 麵粉及梳打粉過篩。

2. 溶解牛油，將它保持溫暖。

3. 將溶牛油、砂糖、黃糖及鹽拌勻，再拌入蛋。

4. 加入粉類，拌勻至剛看不到乾粉為止。注意不要攪拌過度，以免曲奇變硬不夠鬆脆。（看圖 1）

5. 混入谷咕果殼碎、合桃及提子乾。放入雪櫃雪藏至少一小時，雪藏一晚更好。（看圖 2-4）

6. 用雪糕挖將麵糊舀在已墊不黏布或矽膠蓆的焗盤上，每個曲奇之間預留約 1 吋距離。（看圖 5）

7. 放入已預熱 170℃ 的焗爐內，焗約 18-20 分鐘至金黃色，體積越大需時越長。（看圖 6）

** 可製約 30 件曲奇

Nibby Nut and Raisin Cookies

Ingredients

115 g unsalted butter

58 g sugar

58 g brown sugar

2 g sea salt

50 g eggs

160 g all-purpose flour

2 g baking soda

38 g cocoa nibs

58 g roasted walnuts

50 g raisins

Method

1. Sift the flour and baking soda together.
2. Melt the butter and keep warm.
3. Combine melted butter with sugar, brown sugar and salt. Mix well. Whisk in the eggs.
4. Add the dry ingredients from step 1. Mix until they are moistened. Do not over mix it. Otherwise the cookies will be hard instead of crunchy. (see picture 1)
5. Stir in cocoa nibs, walnuts and raisins. Chill for one hour or overnight. (see pictures 2-4)
6. Scoop the dough onto the baking tray. Leave an inch of space between the small dough balls.(see picture 5)
7. Bake in a preheated oven at 170°C for about 18-20 minutes until golden. For larger dough balls, bake for longer time. (see picture 6)

*Makes 30 cookies

白酒燴梨批
White Wine Poached Pear Puff

麥酥皮、白酒燴梨、杏仁忌廉皆是甜品中常見的配搭。麥酥皮香酥，白酒燴梨清香，杏仁忌廉則是各式批撻的基本奶油，三者各具魅力，妙用無窮。

筆者不喜歡在盛暑製作麥酥皮，天氣清涼才能提起勁去擀，一弄就弄多些儲存在冰格，甚麼時候想吃就拿出來烘，配上任何鮮果或奶油都讓人樂而忘憂。

白酒燴梨[1]
White Wine Poached Pear

材料

白酒、水	各 250 克
砂糖，蜂蜜	各 40 克
肉桂	半枝
橙皮、檸檬皮	各半個
檸檬汁	1/4 個
半生啤梨	2 個

做法

1. 除了啤梨外，所有材料煮滾；放進啤梨，轉慢火煮約 15 分鐘。用鐵籤測試，若啤梨外面變軟但裏面稍硬便成。（看圖 1-2）

2. 鋪上牛油紙，讓浮在面的啤梨也能浸於糖水裏。將啤梨浸於糖水裏一晚，使它入味。（看圖 3）

麥酥皮[2]
Puff Pastry

材料

低筋麵粉（建議用日本麵粉）	150 克	鹽	6 克
高筋麵粉（建議用日本麵粉）	150 克	檸檬汁[3]	半個
水	150 克	無鹽牛油（裹入用）[4]	180 克
無鹽牛油（溶化）	30 克	糖霜（灑在餅面用）	適量

做法

1. 將低筋麵粉與高筋麵粉混合，中間開洞，注入水、鹽、檸檬汁及溶化牛油。

2. 用湯匙拌勻，混合成粗糙的麵糰。❺麵糰中間用刀劃一十字，用保鮮膜包好，放進雪櫃約半小時使它鬆弛。（看圖4-5）

3. 無鹽牛油用保鮮膜或膠袋包著，用擀麵棒拍打至正方形，放雪櫃保溫。（看圖6-7）

4. 工作枱撒上足夠手粉。❻用擀麵棒將麵糰從中央向外擀成十字形，中間包入無鹽牛油。（看圖8-11）

5. 壓平麵糰，用擀麵棒擀成長方形，把長度分成四段，左右各向內摺，然後對摺。包上保鮮膜雪藏約半小時，令麵糰鬆弛。（看圖12-16）

6. 重複（5）兩次，摺疊第三次後雪藏約一小時，讓麵糰充分鬆弛。

7. 麵糰擀平，雪約半小時後切出所需形狀，再將做形餅皮放進雪櫃內雪約半小時定形，戳上小孔。（看圖17-18）

杏仁忌廉 ❼

Frangipane Filling (Almond Filling)

材料

杏仁粉	50 克
砂糖	30 克
無鹽牛油	50 克
雞蛋	50 克
低筋麵粉	3 克
冧酒	5 克

做法

1. 無鹽牛油和砂糖混合打發，逐少加入雞蛋打勻。（看圖 19）

2. 輕輕拌入杏仁粉、低筋麵粉及冧酒。（看圖 20）

裝組

1. 從雪櫃取出餅皮，切成 10 厘米 x 10 厘米方塊，塗上蛋漿。（看圖 21-22）

2. 擠上杏仁忌廉，放上半個已切成扇形的白酒燴梨，再灑上糖粉。（看圖 23-28）

3. 放進已預熱 200℃ 的焗爐內，焗約 30 分鐘至金黃色，在啤梨上塗上鏡面果膠。（看圖 29）

** 可做 6 件 10 厘米 x 10 厘米的批餅

貼士

1. 白酒燴梨要預先一天炮製,梨子才會入味,如趕時間,可用罐頭梨代替。宜選用較生的荷蘭啤梨,取其形態優美和果肉爽口。

2. 麥酥皮可預先製作,麵糰可存放冰格達一個月。使用時,在室溫解凍至手指能按下便可使用。

3. 檸檬汁可減少因筋粉與水混合而產生的筋性。

4. 市面上有些高溶點的酥皮油出售,但筆者覺得這種酥油味道古怪,雖然較易操作,但成品不太鬆脆。要成品鬆脆,宜使用真牛油。裹入麵糰前需測試牛油的彈性,如果用手拿著已拍好的牛油,彎曲而不溶化,亦不會因過硬而折斷便可使用。

5. 將麵粉混合成粗糙的麵糰即可,因為麵糰還需擀薄和摺疊三次,麵糰會越來越滑;如此也能減低筋性,方便操作,所以不需要一開始便把麵糰搓滑。

6. 緊記每次摺疊麵糰後要放入雪櫃令麵糰鬆身,而每次擀麵糰時要查看麵糰底部及工作枱上是否有足夠手粉,否則,麵糰會黏著工作枱而弄穿麵糰。麵糰弄穿了,如果是小孔洞是可以用手粉補在上面,但大孔洞就很難補救,勉強補救成品雖然一樣可入爐烘出來,但層次就會因黏著的部分而沒法一層層平均地升起。

7. 杏仁忌廉做法簡單,是法式點心常用的餡料,但切記要加入冧酒才能帶出美味。

White Wine Poached Pear Puff

Production workflow

1. Poach pears in white wine.
2. Make the puff pastry.
3. Make the frangipane filling.
4. Assemble and bake.

White Wine Poached Pear ❶

Ingredients

250 g white wine

250 g water

40 g sugar

40 g honey

1/2 stick cinnamon stick

peel of 1/2 orange

peel of 1/2 lemon

juice of 1/4 lemon

2 half ripened pears

Method

1. Put all ingredients into a pot (except the pears). Bring to the boil. Put in the pears and turn to low heat. Simmer for 15 minutes. Test their doneness with a skewer. If they are soft on the outside but still firm on the inside, they are done. Turn off the heat. (see pictures 1-2)

2. Leave the pears in the syrup. Cover the pears with a sheet of parchment paper. Soak the paper in the syrup so that the pears will be covered in syrup even if they float. Leave them overnight for the flavours to infuse. (see picture 3)

Puff Pastry ❷

Ingredients

150 g cake flour (preferably Japanese)

150 g bread flour (preferably Japanese)

150 ml water

30 g melted unsalted butter

6 g salt

1/2 lemon (juice) ❸

180 g chilled unsalted butter ❹ (for wrapping)

icing sugar (for dusting on top)

Method

1. Mix cake flour and bread flour together on the counter. Make a well at the centre. Pour water, salt, lemon juice and melted butter into the well.

2. Push the flour into the well with a spoon. Then mix well into coarse dough ball. ❺ Make a crisscross cut at the centre of the dough ball. Wrap it in cling film. Rest it in the fridge for 30 minutes. (see pictures 4-5)

3. Wrap the chilled unsalted butter in cling film or a plastic bag. Tap with a rolling pin into a flat square. Refrigerate for later use. (see pictures 6-7)

4. Dust the counter with bread flour ❻ generously. Roll the four quarters of the dough from the cut outward with a rolling pin. The dough should form a cross shape. Put the chilled butter at the centre. Fold the four flaps of dough inward to wrap the butter well. (see pictures 8-11)

5. Flatten the dough. Roll it into a rectangle with a rolling pin. Divide the dough into four quarters across the length. Fold the two end quarters towards the centre. Then fold along the centre line in half. Wrap the dough in cling film and let it rest in the fridge for 30 minutes. (see pictures 12-16)

6. Repeat step 5 twice. Then refrigerate for at last 1 hour so that the dough rests sufficiently.

7. Roll the dough flat. Refrigerate for 30 minutes. Then cut into desired shapes. Leave the cut dough in the fridge for 30 more minutes. Prick holes on the dough with a fork. (see pictures 17-18)

Frangipane Filling (Almond Filling) ❼

Ingredients

50 g ground almonds

30 g sugar

50 g unsalted butter

50 g egg

3 g cake flour

5 g rum

Method

1. Beat the unsalted butter with sugar until light. Pour in the egg a little at a time. Beat well after each addition.

2. Gently stir in ground almonds, cake flour and rum.

Assembly

1. Take the puff pastry dough out of the fridge. Cut into 10cm X 10cm pieces. Brush egg wash over it. (see pictures 21-22)

2. Pipe some frangipane on a piece of dough. Cut a white wine poached pear into halves. Then slice the half pear and fan it out. Arrange on the dough. Sprinkle icing sugar on top. Repeat this step with the remaining dough and pears. (see pictures 23-28)

3. Bake in a preheated oven at 200°C for 30 minutes until golden. Brush pectin glaze over the pears. (see picture 29)

** Makes six 10 cm X 10 cm Pear Puffs.

Tips

1. Poach the pears one day ahead so that the pears have enough time to pick up the flavour. If you're pressed for time, use canned pear instead. Try to get under-ripened Conference pears from the Netherlands, for their curvaceous shapes and crunchy flesh.

2. You may make the puff pastry in advance. The puff pastry can last in the freezer for 1 month. Just thaw the frozen pastry at room temperature until it yields when you press it with finger. Then it is ready to be baked.

3. Adding lemon juice to puff pastry dough helps reduce the gluten formation after the flour comes in touch with water.

4. Puff pastry margarine with higher melting point is available in the market. I've tried it but I personally think it tastes weird. Puff pastry margarine is absolutely easier to handle than real butter in the puff pastry dough, but the pastry ends up not as crispy as one made with real butter. So, I still prefer butter. Just tap the butter into the right size. Check the consistency of the butter before wrapping it in the dough. When you gently bend the butter with your bare hands, it should fold well yet without melting. If the butter snaps right away, it is too cold and too hard.

5. When you mix the ingredients of the puff pastry dough, just roughly mix them into coarse dough. As you have to keep on rolling and folding the dough 3 times, the dough will be smooth as time goes by. You can also reduce the chance of over-kneading the dough this way, and hence reduce the gluten formation. Thus, don't knead the dough smooth right from the start.

6. Make sure you rest the dough in the fridge after each folding. Always check if you have dusted the counter and your hands with enough flour. Otherwise, the dough would stick to the counter and tear the dough apart.

Even if there are holes in the dough, it's not the end of the world. For small holes, simply patch them with some flour. For large holes, there is no perfect way to save them. Just may cut a piece from the corner to patch a hole. The dough can still be baked successfully, but the patched part will not be as flaky and uniformly raised as the rest of the pastry.

7. The frangipane filling is easy to make and it is a common filling for French pastries. Make sure you add some rum to bring out the authentic taste.

糖漬甘橘牛油蛋糕
Kumquat Butter Cake

以牛油、糖、麵粉、蛋為基本材料，日本稱這種從焗爐誕生、
變化無窮的舶來品糕點為燒菓子；它的原產地是法國，則稱之
為 Travelling Cake。加入了清香的甘橘，平衡了糖油的豐盈，
在柑橘當造的季節，豈能錯過。

糖漬甘橘 ❶
Kumquat Compote

材料

新鮮柑橘	200 克
砂糖	150 克
水	200 克
檸檬汁	適量

做法

1. 柑橘洗淨，去蒂和籽後切片（看圖 1）。

2. 砂糖、水煮滾，加入柑橘、檸檬汁，以中火煮約
   15 分鐘，離火，室溫放一晚備用。（看圖 2）。

杏仁牛油蛋糕 ❷
Butter Cake

材料

糖漬甘橘	120 克
無鹽牛油	100 克
糖霜	120 克
雞蛋	180 克
杏仁粉	100 克
低筋麵粉	70 克
發粉	1 克
糖漬甘橘 (裝飾用)	適量
杏仁角 (裝飾用)	適量

做法

1. 麵粉和發粉過篩，加入杏仁粉。

2. 牛油、糖霜打至鬆軟，逐少加入雞蛋拌勻，然後拌入做法 (1)。切忌攪拌過度（看圖 3-4）。

3. 加入已切碎的糖漬甘橘，拌勻。

4. 將麵糊舀進擠花袋，擠進蛋糕模，飾上糖漬甘橘及杏仁角（看圖 5-7）。

5. 放進已預熱 180℃ 的焗爐內，焗約 25 分鐘。將蛋糕移離焗模，噴少許糖水和塗上杏桃果醬，以保濕及增添光澤（看圖 8-9）。❸

** 可做 5 個 12 厘米 x 5.5 厘米 x 5 厘米的小蛋糕

貼士

❶ 預先一天製作糖漬甘橘。甘橘是冬季常見的水果，以日本產的為佳。用糖漬後可耐存放，如放冰格，可保存約一星期。

❷ 製作蛋糕體時，加入麵粉後切勿過度攪拌，應用膠刮輕輕切拌，至不見粉狀物時就可停止。過度攪拌麵粉就會起筋，蛋糕就不鬆軟了。

❸ 燒果子做好不要即日食用，放入密封容器一至兩晚，待回油後食用，風味更佳。

Kumquat Butter Cake

Production worhflow

1. Make the kumquat compote.
2. Make the butter cake.

Kumquat Compote ❶

Ingredients

200 g fresh kumquats
150 g sugar
200 g water
lemon juice as needed

Method

1. Wash and dry the kumquats. Remove stems and seeds. Slice them. (see picture 1)
2. Bring sugar and water to the boil in a pot. Add sliced kumquats and lemon juice. Cook over medium heat for about 15 minutes. Remove from heat. Leave them overnight at room temperature. (see picture 2)

Butter Cake ❷

Ingredients

120 g kumquat compote
100 g butter
120 g icing sugar
180 g eggs
100 g ground almonds
70 g flour
1 g baking powder
kumquat compote (as garnish)
diced almonds (as garnish)

Method

1. Sift flour and baking powder together. Combine with ground almonds.
2. Beat butter and icing sugar together until light and fluffy. Pour in the eggs a little at a time. Mix well after each addition. Fold in flour and almond mixture. Do not over-stir it. (see pictures 3-4)
3. Add kumquat compote and mix well.
4. Put the batter into a piping bag. Pipe the batter into mini loaf tins. Garnish with kumquat compote and diced almonds. (see pictures 5-7)
5. Bake in a preheated oven at 180°C for 25 minutes. Remove from the oven and unmould. Spray some simple syrup (water and sugar) and spread some apricot jam on top for extra sheen and moisture. ❸

(see pictures 8-9)

** Makes five mini loaves measuring 12 cm x 5.5 cm x 5 cm each

Tips

❶ Make the kumquat compote a day ahead. Kumquat is a common winter fruit and those from Japan are the best. The compote can last well in the freezer for a week.

❷ When you make the batter, do not over-stir it after adding flour. Gently fold with a plastic spatula until no dry ingredient is visible. Over-stirring the batter would generate gluten, making the cake chewy and tough instead of fluffy.

❸ Butter cakes are not supposed to be served right away. Keep them in an airtight container for a day or two. They taste even better when given time for the butter to seep back into the cake.

香蕉焦糖燉蛋
Banana Crème Brulee

這道甜點用上冷凍哥斯達黎加香蕉果茸製作，雖然價格較一般新鮮香蕉貴，但只需淺嘗一口，嘴裏的南美風情像在唱：物超所值！

香蕉焦糖燉蛋
Banana Créme Brulee

材料

淡忌廉	240 克
日本牛奶	80 克
冷凍哥斯達黎加香蕉果茸 ❶	200 克
砂糖	36 克
蛋黃	160 克
全蛋	60 克
鹽	1 克
雲呢拿豆莢	1 枝 (用小刀刮出雲呢拿籽)
砂糖或黑蔗糖 (裝飾用)	1 湯匙

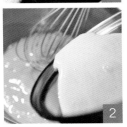
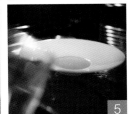

做法

1. 淡忌廉、牛奶、香蕉果茸、雲呢拿豆莢和鹽一同煮滾。離火，讓雲呢拿豆莢浸泡約 10-15 分鐘，釋出香味，棄去雲呢拿豆莢。（看圖 1）

2. 砂糖、蛋黃及全蛋拌勻。

3. 將（1）倒入（2）後拌勻，過篩。（看圖 2-3）

4. 將混合料平均倒入 6-7 隻碟內，每份約 100 毫升，放入已預熱 130℃ 的焗爐，坐熱水焗約 20-25 分鐘。❷（看圖 4-5）

5. 待涼，放進雪櫃冷藏。

燒焦糖香蕉（裝飾用）
Grilled Baby Bananas (as garnish)

材料

迷你香蕉	數隻
砂糖	適量

做法

將香蕉切片，灑上砂糖，以
火槍燒至金黃色。（看圖6）

裝組

1. 食用前，將砂糖薄薄撒於已焗好的燉蛋上，以
   火槍燒成焦糖❸，放入雪櫃雪約 5 分鐘待焦糖
   變硬。（看圖 7-8）

2. 擠上少許加入少量糖打起的淡忌廉，放數片燒
   香蕉和糖水漬芹菜絲❹裝飾。（看圖 9-10）

** 可做約 6-7 碟

貼士

❶ 如果買不到冷凍哥斯達黎加香蕉果
茸，也務必使用優質和熟透的香
蕉。

❷ 焦糖燉蛋要注意燉的時間不可過
火，須比預定時間稍早測試一下是
否烘好。方法是用手指輕輕快手點
一下燉蛋中央，如中央動盪而周圍
凝固就表示烤好。

❸ 完美的焦糖燉蛋在用匙羹舀起來是
呈半凝固狀的，除了要質感軟滑，
有蛋香又有忌廉的細滑之外，最重
要的是燉蛋上那層晶瑩的焦糖脆面
要燒得香脆，最好在食用前才燒
好，燒好的焦糖燉蛋要一小時內享
用，否則焦糖就會潮掉。

❹ 芹菜絲用 1:1 涼糖水浸軟即可。

Banana Crème Brulee

Production workflow

1. Make the banana crème brulee.
2. Grill the baby bananas.
3. Assemble.

Banana Crème Brulee

Ingredients

240 g whipping cream

80 g Japanese whole milk

200 g frozen Costa Rican banana puree ❶

36 g sugar

160 g egg yolks

60 g egg

1 g salt

1 vanilla pod (cut open; seeds scraped off
  with a knife)

1 tbsp sugar or brown sugar (as topping)

Method

1. Bring whipping cream, milk, banana puree,
   vanilla seeds, vanilla pod and salt to the
   boil. Remove from heat. Let the vanilla pod
   to infuse for about 10-15 minutes. Discard
   the vanilla pod. (see picture 1)

2. Whisk sugar, egg yolks and whole egg
   together.

3. Pour the banana mixture from step (1) into
   the egg yolk mixture from step (2). Whisk
   and sift the mixture. (see pictures 2-3)

4. Divide among six to seven 100-ml
   ramekins. Bake in a water bath in a
   preheat oven at 130°C for about 20-
   25 minutes, depending on size of the
   ramekins. ❷ (see pictures 4-5)

5. Leave it to cool. Refrigerate.

Grilled Baby Bananas (as garnish)

Ingredients

baby bananas

sugar

Method

Slice the banana. Sprinkle sugar on top. Burn
the sugar with a propane torch until golden.
(see picture 6)

Assembly

1. Before serving, sprinkle a thin layer of
   sugar on the chilled custard. Burn sugar
   with a propane torch until caramelized. ❸
   Refrigerate for 5 minutes until the caramel
   sets. (see pictures 7-8)

2. Pipe some whipped and sweetened cream
   over the caramel crust. Add a few slices
   of grilled baby banana and garnish with
   candied shredded celery. ❹ (see pictures
   9-10)

** Makes 6 to 7 servings.

Tips

❶ If you can't find Costa Rican banana
  puree, at least get quality and ripened
  bananas.

❷ Do not overcook the custard. In fact,
  you should check its doneness before
  the cooking time is over. To do so, press
  the centre of the custard with your finger
  gently. If it's wobbly and jiggling at the
  centre, but fully set on the rim, the custard
  is done.

❸ The perfect crème brulee should be of
  semi-solid consistency with silky smooth
  texture, strong eggy and creamy flavour.
  The highlight of this dessert lies in the
  crispy caramel crust in amber colour. It
  is the best if you can burn the sugar right
  before you serve. Never leave the caramel
  crust on the custard for more then 1 hour.
  Otherwise, the caramel crust will be soggy,
  sticky and watery instead of crispy.

❹ To make candied celery, shred a celery
  stalk. Then soak it in simple syrup made
  with 1 part of sugar and 1 part of water.
  Soak until the shredded celery is soft.

薫衣草蜂蜜馬卡龍
Lavender Honey Macaroon

做馬卡龍有很多方法，筆者在阿軒師兄的教導下，學了這種較易成功的秘方。其中一個訣竅，是最好使用原粒去皮杏仁，並先把混合好的杏仁粉和糖霜用 bamix 或食物處理器攪成粉末狀。

製作次序		
	1.	馬卡龍
	2.	薰衣草蜂蜜忌廉
	3.	裝組

馬卡龍 ❶

Macaroon

材料

原粒去皮杏仁或杏仁粉	125 克
糖霜	200 克
蛋白	100 克
糖粉	30 克
紫色食物色素	適量

做法

1. 杏仁或杏仁粉與糖霜混合，研磨至細末狀。

2. 蛋白及糖粉一起打發至非常輕，接近發泡膠狀；加入色素。（看圖 1）

3. 拌入杏仁粉混合物，以膠刮片由下而上將麵糊輕輕像折疊起來般，由底部覆上面，直至待麵糊跌落時成三角形，稍為流動即可。（看圖 2-4）

4. 將麵糊舀進擠花袋內，在矽膠蓆上擠 12-14 個 10 厘米直徑的麵糊。（看圖 5）

5. 勿用手指壓扁尖角，輕拍焗盤底部讓麵糊尖角瀉平。

6. 待 5-10 分鐘後，放入已預熱 120℃ 的焗爐內，焗 25 分鐘（烘焗時間視乎麵糊大小而定）。（看圖 6）

薰衣草蜂蜜忌廉 ❷

Lavender Honey Crémeux

材料

薰衣草蜂蜜	50 克
乾燥薰衣草	8 克
蛋黃	60 克
牛奶	80 克
淡忌廉	150 克
魚膠片（用凍水浸軟）	4 克

做法

1. 牛奶及淡忌廉煮熱，加入乾燥薰衣草浸泡 15 分鐘。

2. 蜂蜜加熱至 120℃，與做法 (1) 調勻。

3. 倒入蛋黃內，拌勻。

4. 隔去雜質，回鍋用慢火煮至濃稠，加入已浸軟魚膠片。

5. 再過篩後冷卻，放雪櫃雪至凝固。

裝組

把薰衣草蜂蜜忌廉擠入馬卡龍內，加上裝飾。

** 可做 6-7 個直徑 10 厘米的馬卡龍

竅門

❶ 烘焙馬卡龍最理想是使用可調節淨熱風的焗爐。不過，即使牌子相同的焗爐，火力也有些微差距，所以爐溫和時間必須自己揣摩。總之要找到一個不易令馬卡龍上色的低溫火力，還要很有耐性地觀察，有需要時轉移焗盤位置。

❷ 買不到薰衣草蜂蜜可改用其他蜂蜜，只要稍加乾燥薰衣草份量即可。

Lavender Honey Macaroon

Production workflow

1. Make the macaroons.
2. Make the lavender honey crémeux.
3. Assemble.

Macaroon ❶

Ingredients

125 g blanched whole almonds or ground almonds

200 g icing sugar

100 g egg whites

30 g caster sugar

30 g purple food colouring

Method

1. Mix the almonds with the icing sugar. Grind into powder in a food processor or a hand blender.

2. In a mixing bowl, beat the egg whites and caster sugar until stiff peaks form. Add food colouring. (see picture 1)

3. Fold in the almond mixture with a plastic spatula. (Folding means reaching the bottom of the batter with a spatula, and then lifting the bottom of the batter to the top of it until well mixed.) The batter should drip and hang at the tip of a spatula like a triangle. (see pictures 2-4)

4. Put the batter into a piping bag. On a lined baking tray, pipe 12 to 14 round patties, each about 10 cm in diameter. (see picture 5)

5. The little point that forms when you pipe the macaroon should melt into the macaroon as it settle. If not, tap the bottom of the baking tray to flatten it. Do not use your finger to rub it off.

6. Leave it at room temperature for 5 to 10 minutes. Bake in a preheated oven at 120°C for 25 minutes. (Baking time depends on the sizes of the macaroons.) (see picture 6)

Lavender Honey Crémeux ❷

Ingredients

50 g lavender honey

8 g dried lavender

60 g egg yolk

80 g milk

150 g whipping cream

4 g gelatine leaf (soaked in cold water until soft; drained)

Method

1. Heat up the milk and the whipping cream. Remove from heat and put in the dried lavender. Leave it to infuse for 15 minutes.

2. In another pot, heat honey up to 120°C. Pour it into the cream mixture from step 1. Mix well.

3. Pour the resulting mixture into the egg yolk. Mix well.

4. Pass it through a sieve. Put it back into the pot and cook over low heat until it thickens. Put in the gelatine leaf.

5. Pass it through a sieve again. Leave it to cool. Refrigerate until set.

Assembly

Pipe some Lavender Honey Crémeux on the flat side of a macaroon. Put another macaroon over it with the flat side down. Garnish.

**Makes six to seven 10-cm macaroons

Tips

❶ The recipe works best with convection oven. But even ovens of the same brand may vary in terms of heat. Thus, it is essential that you get to know your own oven well and adjust the baking temperature and time for yourself. Anyhow, the macaroon should not be browned at all. So pick a low temperature to bake it slowly. Observe how the macaroons behave and move the baking tray around if needed.

❷ If you can't get lavender honey, you may use regular one. Just use a little more dried lavender.

熱情果千層酥
Red Fruit Mille Feuilles with Passionfruit Custard

極香、極脆的麥酥皮，夾上酸酸甜甜的熱情果吉士牛油忌廉與鮮雜莓——沒錯，是很高卡路里，沒錯，是難做，是繁複——但當它在舌尖碎成千萬片，它直接了當地告訴你「豈能戒甜」的原因。

製作次序		
	1.	麥酥皮
	2.	熱情果吉士牛油忌廉
	3.	裝組

麥酥皮
Puff Pastry

材料

份量及製作方法
可參考第 95 頁的白酒燴梨批

烘焙方法

1. 烤焙餅皮前，先用一金屬網架放在餅皮上控制餅皮升高的高度。（看圖 1）

2. 以 200℃ 爐溫焗至約 20 分鐘後，翻轉餅皮。如餅皮起泡，可用焗盤壓平餅皮。（看圖 2）

3. 篩上糖粉，餅皮上再放上金屬網架，再放進焗爐烤焗，此時要留意爐溫，勿燒焦糖霜。（看圖 3）

4. 焗至糖粉溶解、金黃，便可出爐。（看圖 4）

熱情果吉士牛油忌廉
Passionfruit Custard Butter Cream

材料

牛奶	170 克
新鮮熱情果汁或冷凍熱情果茸 ❶	40 克
蛋黃	40 克
砂糖	45 克
粟粉	20 克
無鹽牛油 (A)（室溫放軟）	50 克
無鹽牛油 (B)（室溫放軟）	50 克

做法

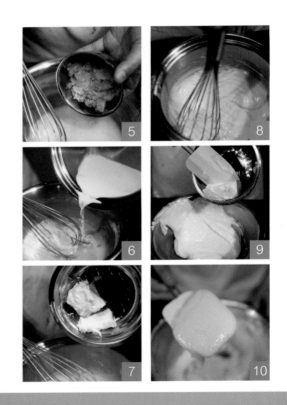

1. 牛奶煮滾。

2. 熱情果茸、蛋黃、粟粉和砂糖拌勻，一邊攪拌一邊慢慢將（1）倒入，拌勻後過篩隔去雜質，回鍋煮滾。（看圖 5-6）

3. 待 (2) 降至微溫，拌入軟無鹽牛油 (A)，放進雪櫃冷藏。（看圖 7-8）

4. 拌入軟無鹽牛油 (B)。❷（看圖 9-10）

裝組

麥酥皮焗好晾涼後切成長條，擠上熱情果吉士牛油忌廉，放上另一片麥酥皮，飾上雜莓。

貼士

❶ 熱情果又稱百香果，果味芬芳馥郁，是多用途的食材。鮮熱情果於每年六月至十二月常見於水果店或市場。以果深紫色、光滑最為新鮮多汁；而皺皮的，水分已經開始流失，但味道較甜。熱情果只需用小刀切開，取果肉，並隔出汁液就可使用。未隔出汁液的果肉，可冷凍數月留用。買不到鮮果，可使用冷凍果茸，使用方便，質量平均。

❷ 牛油分兩次加入，可使忌廉更順滑、輕盈。

Red Fruit Mille Feuilles with Passionfruit Custard

Production workflow

1. Make the puff pastry. Bake it.
2. Make the passionfruit custard butter cream.
3. Assemble.

Puff Pastry

Ingredients and method

Please refer to the recipe puff pastry with white wine poached pear on p.96 for ingredients and method of the dough.

Baking Method

1. In the beginning, place another wire rack over the puff pastry so as to control its height. (see picture 1)
2. Bake at 200°C for 20 minutes. Flip the pastry over. If it warps or blisters, press it flat with a baking tray. (see picture 2)
3. Sprinkle icing sugar on the pastry. Place wire rack over the puff pastry. Bake in the oven again. Control the temperature carefully so as not to burn the icing sugar. (see picture 3)
4. Bake until the icing sugar melts and golden. Set aside to let cool. (see picture 4)

Passionfruit Custard Butter Cream

Ingredients

170 g milk
40 g fresh or frozen passionfruit puree ❶
40 g egg yolk
45 g sugar
20 g cornflour
50 g unsalted butter (A) (at room temperature)
50 g unsalted butter (B) (at room temperature)

Method

1. Bring the milk to the boil.
2. Mix passionfruit puree, egg yolk, cornflour and sugar well. Pour the boiled milk into egg yolk mixture, a little at a time. Sift the mixture. Cook until it boils. (see pictures 5-6)
3. Let it cool to lukewarm. Fold in butter (A). Refrigerate. (see pictures 7-8)
4. Stir in the soft butter (B). ❷ (see pictures 9-10)

Assembly

Cut the puff pastry into long strips. Pipe passionfruit custard butter cream on it. Top with another strip of puff pastry. Garnish with mixed berries. Serve.

Tips

❶ Passionfruit has a unique fruity flavour that makes it a versatile food ingredient. Fresh passionfruit are available in fruit stalls or the market from June to December every year. For culinary use, pick those with smooth deep purple skin for freshness and juiciness. Those with wrinkly skin are less juicy but they tend to taste sweeter which means you may serve them straight as a fruit. To get the pulp, simply cut open a passionfruit. Then pass it through a sieve to get the juice. Generally speaking, passionfruit pulp with seeds in it can be frozen and last for months. If you can't get the fresh fruit, use frozen puree instead for similar taste and result.

❷ Adding half of the butter into the custard butter cream at one time makes it lighter and smoother.

無花果朱古力批
Fresh Fig Chocolate Flute Tart

無花果清甜得像玉骨冰肌，煮成果醬，配上吉士忌廉，反變得幽香豐
潤。把它們滿滿地鋪在朱古力鬆脆餅底上，佐以輕盈的英國早餐茶，
最是愜意。

製作次序	1. 朱古力鬆脆餅底	4. 裝組
	2. 無花果果醬	
	3. 吉士忌廉	

朱古力鬆脆餅底 ❶
Chocolate Pâte Sablée Shortcrust

材料

低筋麵粉	110 克
高筋麵粉	40 克
可可粉	12 克
發粉	10 克
無鹽牛油 （室溫放軟）	110 克
砂糖	100 克
黑朱古力（50％以上谷咕含量）(切碎)	50 克
蛋黃	50 克
黑朱古力 (溶化，掃面用)	適量

做法

1. 麵粉、可可粉、發粉一同過篩。

2. 朱古力坐熱水或用微波爐加熱至溶化。（看圖 1）

3. 牛油及砂糖拌勻，直至砂糖溶化。

4. 將蛋黃逐少加入做法 (3) 內，再加入溶化的朱古力，拌勻。（看圖 2）

5. 輕輕拌入粉類，不要攪拌過度以至成品過硬。用保鮮紙包裹，放入雪櫃雪藏約半小時。（看圖 3）

6. 將麵糰壓在直徑 20 厘米的餅模裏至八成滿，放上矽膠蓆、再壓一焗盤及可入爐的重物，放入已預熱 170℃ 的焗爐內，焗約 25 分鐘。略放涼後倒扣脫模，直至涼透。（看圖 4-6）

無花果果醬
Fig Compote

材料

冷凍或新鮮無花果 ❷	450 克
砂糖	150 克
果膠	5 克
檸檬汁	20 克

做法

1. 少量砂糖與果膠混合。❸

2. 其餘砂糖和無花果用小火煮至水分開始收乾，下果膠糖，繼續煮至濃稠，下檸檬汁調味，離火，放涼待用。(看圖 7-9)

吉士忌廉
Custard Cream

材料

牛奶	170 克
淡忌廉	40 克
蛋黃	40 克
砂糖	45 克
粟粉	20 克
無鹽牛油（室溫放軟）	50 克
雲呢拿豆莢	半枝（用小刀刮出雲呢拿籽）

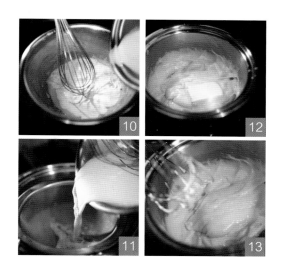

做法

1. 將雲呢拿籽連同豆莢放進牛奶、忌廉裏，煮滾，關火。

2. 蛋黃、粟粉及砂糖拌勻，一面攪拌一面加入做法（1），回鍋，一邊以細火加熱一邊攪拌，直至沸騰，離火過篩，待涼。（看圖 10-11）

3. 拌入軟牛油，放入雪櫃備用。（看圖 12-13）

裝飾

新鮮無花果（切片）適量

裝組

1. 在餅底上掃上溶朱古力，雪一會至朱古力凝固。（看圖 14）

2. 塗一層吉士忌廉，一層無花果果醬。（看圖 15-17）

3. 排上新鮮無花果，最後在鮮果上塗上鏡面膠，以保持新鮮。（看圖 18-19）

** 可做 1 個 20 厘米果撻

竅門

1 在製作朱古力鬆脆餅底時留意勿過度攪拌，麵糰一定要放入雪櫃雪上最少半小時才入模。因為食譜內有發粉，填入模只需約八成的份量，再用不黏布或矽膠蓆蓋面，上面壓一焗盤和重物，這樣做可令批底平滑。焗後不要立即將餅底反扣，須晾涼一會，然後才倒扣在網架上繼續待涼。

2 新鮮無花果以以色列出產的為佳，果內紅色果蕊越多越清甜。

3 果膠是從水果提鍊的膠質，呈啫喱粉狀，是一種增厚劑。必須與少許砂糖混合方加入水果內熬煮，否則會容易結粒。市面有一些已混合果膠的砂糖出售，但價錢較貴，可向 DIY 店舖購買散裝果膠。

Fresh Fig Chocolate Flute Tart

Production workflow
1. Make the chocolate pâte sablée shortcrust.
2. Make the fig compote.
3. Make the custard cream.
4. Assemble the tart. Garnish with fresh fig.

Chocolate Pâte Sablée Shortcrust **1**
Ingredients
110 g cake flour

40 g bread flour

12 g cocoa powder

10 g baking powder

110 g unsalted butter (at room temperature)

100 g sugar

50 g dark chocolate (over 50% cocoa) (finely chopped)

50 g egg yolk

molten chocolate (for brushing)

Method
1. Sift the flour, cocoa powder and baking powder together.
2. Melt the chocolate over a hot water bath or in a microwave. (see picture 1)
3. Blend the butter with sugar until the sugar moistened by the butter.
4. Add a little egg yolk at a time. Mix after each addition. Add the melted chocolate and mix well. (see picture 2)
5. Fold in dry ingredients from step 1. Do not over-stir. Otherwise the pastry will be too hard. Wrap the dough with cling film. Chill the dough for 30 minutes. (see picture 3)
6. Press the dough into the tart mould up to 80% of the height of the mould. Line the dough with silicone mat and put another tart mould and heavy objects on top. Blind bake in a preheated oven at 170°C for about 25 minutes until golden. Remove from the oven. Unmould when the pastry has cooled slightly. Put the unmoulded pastry on a wire rack until completely cool. (see pictures 4-6)

Fig Compote
Ingredients
450 g frozen or fresh figs **2**

150 g sugar

5 g pectin

20 g lemon juice

Method

1. Mix a little sugar with pectin.❸ Set aside.

2. Cook the remaining sugar and figs over low heat until the liquid reduces. Add the pectin-sugar mixture from step 1. Keep on cooking until thick. Add lemon juice and stir well. Remove from heat. Leave it to cool. (see pictures 7-9)

Custard Cream

Ingredients

170 g milk

40 g whipping cream

40 g egg yolk

45 g sugar

20 g cornflour

50 g unsalted butter (at room temperature)

1/2 vanilla pod (split open; seeds scraped off with a knife)

Method

1. In a pot, put in the vanilla pod, vanilla seeds, milk and whipping cream. Bring to the boil. Turn off the heat.

2. In a mixing bowl, whisk the egg yolk, cornflour and sugar together. Pour the hot cream mixture from step 1 into this egg yolk mixture while stirring continuously. Pour the resulting mixture back into the pot. Put it over low heat while stirring constantly until it boils. Remove from heat and pass it through a sieve. Leave it to cool. (see pictures 10-11)

3. Stir in the butter. Refrigerate for later use. (see picture 12-13)

Garnish

fresh figs (sliced)

Assembly

1. Brush molten chocolate on the inside of the cooled pastry. Refrigerate until set. (see picture 14)

2. Spread a layer of custard cream in the tart. Top with a layer of fig compote. (see pictures 15-17)

3. Arrange sliced fresh figs on top. Brush a layer of pectin glaze over the figs to shine them up and keep them fresh. (see pictures 18-19)

**Makes one 20-cm round tart.

Tips

❶ Do not over-mix the dough when you make the chocolate pâte sablée shortcrust. Leave the dough to rest in the fridge for at least 30 minutes before pressing it into the tart mould. As there is baking powder in the dough, the dough should reach only 80% of the height of the mould. Lining the tart pastry with non-stick baking lining or silicone mat, on top of which a baking tray and some oven safe heavy object are added. This step ensures the tart will end up smooth and even in thickness. Do not turn the tart pastry out immediately after baking. Leave it to cool for a while and firm up a little before turning it out and cool it on a wire rack.

❷ The best fresh figs are from Israel. The more pinkish red pulp it has inside, the sweeter the fig is.

❸ Pectin is a gelatinous substance extracted from fruit. It is available in powder form and used as a gelling agent. Pectin should be mixed with a little sugar before added to the hot fruit mixture. Otherwise, it might get lumpy. In the market there is pectin powder premixed with sugar. But it is sold at the higher price than pectin powder alone. Get pectin powder in bulk from DIY baking supply stores.

藍色歌劇院
Blue Opera

把傳統的咖啡味 Opera 變奏成藍調——富含果味卻又香滑濃重，配以杏仁薄蛋糕、藍莓糖漿、朱古力乳霜，不同層次的紫色，口味清新而優雅。

傳統的 opera，高度需在 2.5-3 厘米之內、7-9 層，做成方形，糖漿要滲滿每層蛋糕，看不到蛋糕體。

製作次序	1. 杏仁薄蛋糕	4. 朱古力乳霜	7. 裝組
	2. 藍莓牛油忌廉	5. 底層朱古力	
	3. 藍莓糖漿	6. 朱古力淋面	

杏仁薄蛋糕 ❶

Almond Joconde

材料

杏仁粉	80 克
糖霜	50 克
全蛋	60 克
蛋黃	40 克
蛋白	150 克
砂糖	60 克
低筋麵粉	75 克

** 可製 1.5 厘米 x 30 厘米 x 40 厘米長方形蛋糕兩片

做法

1. 糖霜與杏仁粉混合。（看圖 1）

2. 全蛋、蛋黃與（1）混合，打發成白色、飽含空氣的鬆軟狀。（看圖 2-3）

3. 蛋白與砂糖打至企身，將一半蛋白糊拌入（2）。（看圖 4-6）

4. 加入低筋麵粉，拌勻，再加入剩餘的蛋白糊，輕輕拌勻。（看圖 7）

5. 倒入墊上矽膠蓆或不黏布的烤盤，抹平約 0.5 厘米厚，以 200℃ 焗 9-10 分鐘，取出，攤涼。將每片蛋糕切成 15 厘米 x 20 厘米的長方片。（看圖 8）

藍莓牛油忌廉 ❷
Blueberry Butter Cream

材料

藍莓果茸	200 克
覆盆子果茸 ❸	50 克
蛋	120 克
砂糖	125 克
意大利蛋白霜	125 克
雲呢拿豆莢	1 枝 (切開豆莢，刮籽)
無鹽牛油（室溫放軟）	380 克

做法

1. 藍莓果茸及紅桑子果茸混合煮滾，關火。

2. 蛋及砂糖拌勻，然後一面攪拌一面慢慢加入已煮滾的（1）❹。（看圖 9）

3. 隔去雜質，隔熱水加熱，不斷攪拌，直至溫度達 85℃。（看圖 10）

4. 隔去雜質。待果茸蛋糊涼卻，加入雲呢拿籽，拌入軟牛油。（看圖 11）

5. 最後拌入意大利蛋白霜，打勻，待用。（看圖 12-13）

** 意大利蛋白霜做法，請參考第 158 頁的柚子撻

藍莓糖漿
Blueberry Syrup

材料

藍莓果茸	100 克
紅桑子果茸	40 克
砂糖	45 克
水	150 克

做法

將全部材料煮熱，關火後待涼。

朱古力乳霜 ❺

Ganache

材料

黑朱古力（50% 谷咕含量以上）（切細）	60 克
淡忌廉	30 克
牛奶	20 克
無鹽牛油（室溫放軟）	10 克

做法

1. 淡忌廉及牛奶煮滾，倒入朱古力，輕輕攪拌至溶化。

2. 加入牛油拌勻，待用。

底層朱古力

Chocolate Layer

材料

黑朱古力（50% 谷咕含量以上）	80 克

做法

1. 朱古力隔熱水輕輕攪拌至溶化。

2. 塗在其中一片杏仁薄蛋糕上，放入雪櫃冷藏，待用。（看圖 14-15）

朱古力淋面

Blueberry White Chocolate Glazing

材料

白朱古力	100 克
淋面白朱古力 ❻	100 克
藍莓果茸	75 克
覆盆子果茸	30 克
葡萄糖膠	25 克

做法

1. 將白朱古力切碎，加入淋面白朱古力。

2. 將藍莓果茸、覆盆子果茸及葡萄糖膠一同煮滾，倒入白朱古力及淋面白朱古力內，攪勻。(看圖 16-17)

3. 過篩，並放涼至 40℃，淋在已冷凍的蛋糕面上。

裝組

1. 預備 1 個 4.5 厘米 x 15 厘米 x 20 厘米長方形模。

2. 這個藍色歌劇院裝組較為複雜，所以我用以下的平面圖和輔以步驟圖給大家跟隨。

平面圖

朱古力淋面
藍莓牛油忌廉
杏仁薄蛋糕塗上藍莓糖漿
朱古力乳霜
杏仁薄蛋糕塗上藍莓糖漿
藍莓牛油忌廉
杏仁薄蛋糕塗上藍莓糖漿
底層朱古力

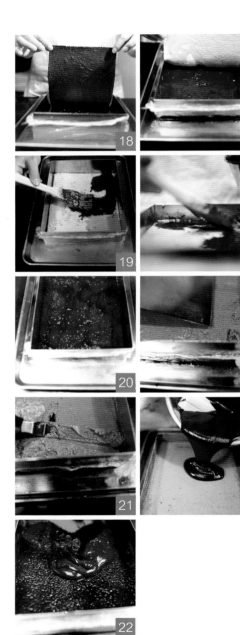

3. 首先在杏仁薄蛋糕塗上底層朱古力，雪凍；裝組時，將朱古力向底、蛋糕向面。蛋糕塗上藍莓糖漿，放入藍莓牛油忌廉，抹平；放上一片杏仁薄蛋糕，塗上藍莓糖漿，倒入朱古力乳霜；再放上一片杏仁薄蛋糕，塗上藍莓糖漿，放入藍莓牛油忌廉，抹平，再淋朱古力淋面，即大力告成。雪凍後享用。(看圖 18-26)

** 可做一個 3 厘米 x 15 厘米 x 20 厘米長方形蛋糕

貼士

❶ 杏仁薄蛋糕 (Biscuit Joconde)，Biscuit 是指用打起的蛋白令其膨脹鬆軟的蛋糕，方法是將蛋黃與蛋白分別打起及加糖，最後加入麵粉，目的是令麵糊內部充滿空氣，烘焗後會變得分外鬆軟。因為蛋糕很薄，烘焙要用比較高溫度，迅速地烘熟，避免在過長時間烘焗後水分給烘乾，蛋糕變得乾韌。

❷ 藍莓牛油忌廉要打得很軟，在塗抹滲滿糖漿的蛋糕體時就會較容易。如想每層牛油忌廉的厚度都一致，可以用磅量好每層所用份量才塗抹，每層約需 150 克 藍莓牛油忌廉

❸ 藍莓的色素當接觸空氣就很不穩定，會由紫藍色慢慢變成灰色，做不到預期的效果，所以必需加入紅色的水果或果茸調色，讓它變成穩定的紫色。

❹ 把煮熱果茸加入蛋黃糖糊內，再加熱至 85℃，是因為以 85℃煮 2-3 分鐘可大大減低蛋黃內的沙門氏菌含量。

❺ 朱古力乳霜是由朱古力和忌廉製成的乳霜，用於做餡料或塗在甜點面層。

❻ 淋面朱古力有黑、白、奶三種，用植物油製造。主要配合調溫朱古力一起使用，溶解後淋在糕點上作裝飾。

Blue Opera

Production workflow

1. Make the almond Joconde.
2. Make the blueberry butter cream.
3. Make the blueberry syrup.
4. Make the ganache.
5. Make the chocolate layer.
6. Make the blueberry white chocolate glazing.
7. Assemble the cake in the rectangular mould. Pour glazing on top.

Almond Joconde❶

Ingredients

80 g ground almonds
50 g icing sugar
60 g egg
40 g egg yolk
150 g egg whites
60 g sugar
75 g cake flour

**Makes two slices measuring 30 cm x 40 cm x 1.5 cm

Method

1. Mix icing sugar and ground almonds together. (see picture 1)
2. Add egg and egg yolk to the almond mixture from step 1. Beat until light and fluffy. (see pictures 2-3)
3. In another bowl, beat egg whites with sugar over medium speed until stiff. Fold half of the egg whites into the egg yolk mixture from step 2. (see pictures 4-6)
4. Add cake flour and mix well. Then fold in the remaining half of egg whites. Mix well. (see picture 7)
5. Pour the batter into baking tray lined with a non-stick baking lining or silicone mat, the batter is about 0.5 cm thick. Bake in a preheated oven at 200°C until golden for 9 to 10 minutes. Leave it to cool. Cut each slice into halves, each measuring 15 cm x 20 cm. (see picture 8)

Blueberry Butter Cream[2]

Ingredients

200 g blueberry puree
50 g raspberry puree [3]
120 g eggs
125 g sugar
125 g Italian meringue
1 vanilla pod (split open; seeds scraped off with a knife)
380 g unsalted butter (at room temperature)

Method

1. Bring the blueberry puree and raspberry puree to the boil. Remove from heat.
2. Mix the eggs and sugar. Pour in boiled puree while stirring continuously.[4] (see picture 9)
3. Pass the mixture through a sieve. Then heat the sifted mixture over a hot water bath up to 85°C. (see picture 10)
4. Sieve the mixture again and leave it to cool. Add vanilla seeds and fold in butter. (see picture 11)
5. Mix with Italian meringue. (see pictures 12-13)

   (refer to Yuzu Tart on p.162 for method)

Blueberry Syrup

Ingredients

100 g blueberry puree
40 g raspberry puree
45 g sugar
150 g water

Method

Bring all the ingredients to the boil. Remove from heat and let cool.

Ganache [5]

Ingredients

60 g dark chocolate (at least 50% cocoa;
finely chopped)
30 g whipping cream
20 g milk
10 g unsalted butter (at room temperature)

Method

1. Bring cream and milk to the boil. Then pour onto the chocolate. Stir until chocolate melts.
2. Add butter and mix well.

Chocolate Layer

80 g dark chocolate (at least 50% cocoa)

Method

1. Melt chocolate over a hot water bath.
2. Spread onto a piece of Almond Joconde. Refrigerate for later use. (see pictures 14-15)

Blueberry White Chocolate Glazing

Ingredients

100 g white chocolate
100 g white coating chocolate [6]
75 g blueberry puree
30 g raspberry puree
25 g glucose syrup

Method

1. Chop the white chocolate into chunks. Mix with white coating chocolate.
2. Bring the blueberry puree, raspberry puree and glucose syrup to the boil. Add chocolate and mix well. (see pictures 16-17)
3. Sieve the mixture. Let it cool to 40°C. Pour over the chilled cake.

Assembly

As this is a rather complicated recipe, I drew

a layered diagram to illustrate the assembly steps:

Chocolate Glazing
Blueberry Butter Cream
Almond Joconde Infuse with Blueberry syrup
Ganache
Almond Joconde Infuse with Blueberry syrup
Blueberry Butter Cream
Almond Joconde Infuse with Blueberry syrup
Melt Chocolate Layer

1. Pour chocolate on one sheet of almond Joconde. Refrigerate until set.
2. Put the chocolate layer to a 15 cm X 20 cm mould with the chocolate side facing down. (see picture 18)
3. Brush blueberry syrup on the Joconde. (see pictures 19-20)
4. Spread a thin layer of blueberry butter cream over the Joconde. Smooth it out. (see picture 21)
5. Top with another layer of almond Joconde. Brush blueberry syrup on it. Spread a layer of ganache over it. (see picture 22)
6. Place another layer of almond Joconde on top. Brush blueberry syrup. Spread a layer of blueberry butter and smooth it out. Pour the blueberry white chocolate glazing on top. (see pictures 23-26)
7. Refrigerate until set. Serve.
** Makes one 3 cm x 15 cm x 20 cm rectangular cake.

Tips

1. Joconde biscuit is a thin almond sponge raised by stiff and foamy egg whites. The egg yolks and whites are separated first and beaten separately. Sugar and flour are then added at last. The air in the stiff egg whites is kept in the batter so that the sponge is fluffy and light after baked. Joconde is very thin and it should be baked at higher temperature for a shorter period of time. Otherwise, the moisture will be evaporated and the sponge becomes chewy.

2. The blueberry butter cream should be beaten to a thin spreadable consistency so that it can be applied easily over the syrup-infused almond Joconde. For the most exquisite presentation, each layer of butter cream should be of the same thickness. To achieve that, weigh the butter cream before applying. Each layer of butter cream should weigh around 150 g.

3. The pigment in blueberries becomes unstable and turns grey slowly once in touch with air. Thus, for a stable and vibrant purple colour, red fruit or fruit puree should be added to the blueberries.

4. When we make the blueberry butter cream, the hot fruit puree is added to the egg yolk-sugar mixture, which is then heated up to 85°C. This step is important as cooking it at 85°C for 2 to 3 minutes significantly reduces the number of salmonella bacteria in the egg yolk.

5. Ganache is a runny glossy glazing made from chocolate and cream. It is commonly used as a filling or as an icing on top of sweet pastries.

6. Coating chocolate is available in dark, milk and white. It is usually mixed with couverture chocolate before melted and poured on cakes as finish.

異國風情
Exotic Delight

熱情未必如火，如冰豈非更暢快？杏仁脆餅、熱帶風味餡心、白芝士慕絲，有如樹影婆娑的南美沙灘上，仙子乘風踏浪而來。

杏仁脆餅 ❶

Almond Shortbread Pastry

材料

無鹽牛油（室溫放軟）	50 克
糖霜	30 克
鹽	1 克
雞蛋	20 克
低筋麵粉	87 克
杏仁粉	12 克
發粉	1 克
烘香杏仁片	26 克
溶化白朱古力	適量

** 預備 1 個 15 厘米 x 20 厘米的長方形模

做法

1. 牛油、糖霜及鹽用膠刮輕輕拌勻，至糖霜被牛油吸收。

2. 逐少加入雞蛋，攪拌至蛋漿被牛油糖糊完全吸收。

3. 加入粉類及烘香杏仁片，拌勻成麵糰。

4. 將麵糰冷藏至硬，用擀麵棒將麵糰略敲扁，再擀薄至約 0.2 厘米，然後切出 15 厘米 x 20 厘米的餅皮，戳上小孔。（看圖 1-2）

5. 連鐵圈放入已預熱 180℃ 的焗爐內，焗約 15-18 分鐘至金黃色。（看圖 3）

6. 待涼後，塗上溶化白朱古力，冷藏待用。（看圖 4）

熱帶風味餡心 ❷
Exotic Jelly Filling

材料

冷凍菠蘿果茸	75 克
冷凍芒果果茸	35 克
冷凍青檸果茸	20 克
砂糖	28 克
水	35 克
椰奶	22 克
椰子白冧酒	11 克
魚膠片 (用凍水浸軟)	6 克

做法

1. 將所有果茸、砂糖及水煮滾，加入已浸軟的魚膠片、椰奶及白冧酒。（看圖 5-7）

2. 倒進長方形模內，放入冰格冷凍至結冰。

** 預備 1 個 15 厘米 x 20 厘米的長方形模

杏仁薄蛋糕
Almond Joconde

材料

杏仁粉	50 克	砂糖	35 克
糖霜	30 克	低筋麵粉	40 克
蛋黃	30 克	無鹽牛油 (溶化、保持溫暖備用)	20 克
全蛋	40 克		
蛋白	90 克		

** 預備 30 厘米 x 40 厘米的焗盤 2 個

做法

1. 請參考第 124 頁的藍色歌劇院內杏仁薄蛋糕的做法（1）至（3）。

2. 加入低筋麵粉，拌匀，再加入剩餘的蛋白糊，輕輕拌匀；舀兩湯匙麵糊和溶牛油拌匀，再倒入其餘麵糊一起拌匀。

3. 倒入墊上矽膠蓆或不黏布的烤盤，抹平約 0.8 厘米厚，以 200℃焗約 10-12 分鐘，取出攤涼，將每片蛋糕切成 15 厘米 x 20 厘米長方片，共 4 片。

白芝士慕絲

Fromage Frais Mousse

材料

砂糖 (A)	5 克	法國白芝士 (40% 脂肪)	220 克
蛋黃	50 克	魚膠片 (用凍水浸軟)	8 克
砂糖 (B)	85 克	打起淡忌廉	250 克
水	25 克	檸檬皮茸	8 克

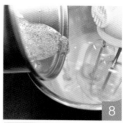

做法

1. 蛋黃與砂糖（A）拌勻。❹

2. 砂糖（B）及水煮至 100℃ 時，開始打發 (1) 至濃稠。待糖水煮至 118℃ 時，把糖水慢速流暢地撞入 (1) 內，繼續高速打發至蛋黃糖糊 變得濃稠。❺（看圖 8）

3. 加入已浸軟及預先溶解的魚膠，過篩，濾走被灼熟結塊的蛋黃和其他 雜質。（看圖 9）

4. 待蛋黃糖糊涼至約室溫，便可一杓一杓地加入白芝士，每加一杓也要 用打蛋器拌勻，才加另一杓。（看圖 10）

5. 加入所有芝士後，分數次加入打起的淡忌廉和檸檬皮茸，輕輕拌勻。❻ （看圖 11-12）

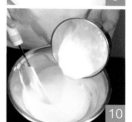

熱帶風味糖漿

Exotic Punch Syrup

材料

冷凍菠蘿果茸	65 克
冷凍芒果果茸	30 克
冷凍青檸果茸	15 克
砂糖	25 克
水	32 克
椰奶	20 克
白椰子冧酒	10 克

做法

將所有果茸、砂糖及水煮滾，離火。加入 椰奶及白冧酒，拌勻。

裝組

1. 將塗了白朱古力的杏仁脆餅放在包了保鮮膜的長方形慕絲框底部，白朱古力面向上。

2. 抆上一層很薄的白芝士慕絲，以防蛋糕出模後移位。放一片杏仁薄蛋糕，塗上熱帶風味糖漿。（看圖 13）

3. 舀一層白芝士慕絲，抆平。

4. 加一片杏仁薄蛋糕，塗上熱帶風味糖漿，再抆一層很薄的白芝士慕絲。（看圖 14-16）

5. 放上熱帶風味餡心。（看圖 17）

6. 抆一層很薄的白芝士慕絲，再放上一片杏仁薄蛋糕，塗上熱帶風味糖漿。

7. 加入餘下的白芝士慕絲，伴上裝飾即成。（看圖 18）

** 可製成 1 個 15 厘米 x 20 厘米 x 4.5 厘米的長方形蛋糕

貼士

❶ 杏仁脆餅主要特點是香脆，為糕餅提供另一種口感。這種脆餅多採用糖霜而不用砂糖，因為糖霜溶於牛油的速度較快，使麵糰較細緻；若採用砂糖，部分糖的晶粒在拌合時未能溶解，但在製作餡餅皮時才溶，形成滲漏，令麵糰較易撕裂、以及容易黏在工作枱及捍麵棒上，這個情況在天氣越熱時越嚴重。要解決問題，便得一再灑上麵粉，最終令麵糰變硬，烘焗後在表面亦會出現一層殼，大大破壞了口感。

要麵糰烘焗後香脆，主要是在搓麵粉時盡量避免麵筋出現，方法是減少麵粉與水分接觸，或者在烘焗前二者的接觸時間越短越好。

❷ 如買不到餡心的果茸材料，可用鮮果汁代替。

❸ 白芝士慕絲採用法國白芝士 (fromage frais Onctueux)，分別有 0% 脂肪含量和 40% 脂肪含量兩種，為使蛋糕更絲滑，宜選用後者。

❹ 製作蛋黃糖糊時，蛋黃內加入小量糖，可以防止加入熱糖漿時煮熟蛋漿而結塊。

❺ 將糖和水煮至約 118℃，再倒進蛋黃裏以高速打發，可殺死大量沙門氏菌，並帶出蛋黃的香味。

❻ 混合淡忌廉時，為免攪拌過度，最好使用膠刮。

Exotic Delight

Production workflow

1. Make the almond shortbread pastry.
2. Make the exotic jelly filling.
3. Make the almond Joconde.
4. Make the fromage frais mousse.
5. Make the exotic punch syrup.
6. Assemble the cake and garnish.

Almond Shortbread Pastry ❶

Ingredients

50 g unsalted butter (at room temperature)

30 g icing sugar

1 g salt

20 g egg

87 g cake flour

12 g ground almond

1 g baking powder

26 g toasted flaked almonds

molten white chocolate

** Prepare a 15 cm x 20 cm rectangular baking tray.

Method

1. Mix butter, icing sugar and salt together with a plastic spatula until the icing sugar is moistened by the butter.
2. Add the egg little by little and stir well after each addition. Stir until well incorporated.
3. Put in the dry ingredients and toasted almonds. Stir into dough.
4. Refrigerate the dough until firm. Bash it gently with a rolling pin to flatten slightly. Roll it out with a rolling pin into a dough, and 0.2 cm thick. Cut into 15 cm x 20 cm in size. Prick holes on the dough sheet with a fork. (see pictures 1-2)
5. Bake in a preheated oven at 180˚C for 15 to 18 minutes until golden. (see picture 3)

6. Leave it to cool. Brush the molten white chocolate over the cooled pastry. Refrigerate for later use. (see picture 4)

Exotic Jelly Filling ❷

Ingredients

75 g frozen pineapple puree

35 g frozen mango puree

20 g frozen lime puree

28 g sugar

35 g water

22 g coconut milk

11 g Malibu rum

6 g gelatine leaf (soaked in cold water until soft; drained)

** Prepare a 15 cm X 20 cm rectanglular baking tray.

Method

1. Bring pineapple puree, mango puree, lime puree, water and sugar to the boil. Add soaked gelatine leaf, coconut milk and Malibu rum. Stir until gelatine dissolves. (see pictures 5-7)
2. Pour the mixture into the oblong mould and freeze it.

Almond Joconde

Ingredients

50 g ground almonds

30 g icing sugar

30 g egg yolk

40 g egg

90 g egg whites

35 g sugar

40 g cake flour

20 g unsalted butter (melted and keep warm)

** Prepare two 30 cm X 40 cm rectangular baking tray.

Method

1. Refer to the recipe of Almond Joconde on p.129 steps 1-3 for method.

2. Add cake flour and mix well. Then fold in the remaining half of egg whites. Add two tbsp of this batter into the melted butter and mix well. Put the butter mixture back in. Mix well.

3. Pour the batter into baking tray lined with a non-stick baking lining or silicone mat, the batter is about 0.8 cm thick. Bake in a preheated oven at 200°C until golden for 10 to 12 minutes. Leave it to cool. Cut each slice into four sheets, each measuring 15 cm X 20 cm.

Fromage Frais Mousse

Ingredients

5 g sugar (A)

50 g egg yolk

85 g sugar (B)

25 g water

220 g fromage frais (40% fat) ❸

8 g gelatine leaf (soaked in cold water until soft; drained)

250 g whipped cream

8 g lemon zest

Method

1. Mix egg yolk with sugar (A). ❹

2. In a pot, heat sugar (B) and water until it reaches 100°C. Start to beat the egg yolk-sugar mixture until thick. When the sugar syrup reaches 118°C, pour the boiling syrup slowly into the egg yolk-sugar mixture. Beat over high speed until it thickens. (see picture 8) ❺

3. Put in the gelatine leaf. Pass the mixture through a sieve to remove any impurity and lump. (see picture 9)

4. Leave the custard to cool. Add one ladle of fromage frais at a time and beat well after each addition. (see picture 10)

5. Then add a little whipped cream at a time. Fold well after each addition. Add lemon zest and fold well. Refrigerate for later use. (see pictures 11-12) ❻

Exotic Punch Syrup

Ingredients

65 g pineapple puree

30 g mango puree

15 g lime puree

25 g sugar

32 g water

20 g coconut milk

10 g Malibu rum

Method

Bring all fruit purees, sugar and water to the boil. Remove from heat. Add coconut milk and Malibu rum. Stir well.

Assembly

1. Wrap the bottom of a rectangular mousse ring with cling film. Press the almond shortbread pastry on the bottom of the ring with white chocolate side facing up.

2. Spread a thin layer of fromage frais mousse over the pastry to secure the shape of the cake after unmoulded. Then put a piece of almond Joconde on top. Brush exotic punch syrup over the Joconde. (see picture 13)

3. Pour a thicker layer of fromage frais mousse on top. Smooth it out.

4. Put on a piece of almond Joconde. Brush on exotic punch syrup. Spread a thin layer of fromage frais mousse. (see pictures 14-16)

5. Place the frozen exotic jelly filling on top. (see picture 17)

6. Spread another thin layer of fromage frais mousse. Top with another piece of almond

Joconde. Brush on exotic punch syrup.

7. Pour in the remaining fromage frais mousse. Garnish. Serve. (see picture 18)

** Make a 15 cm x 20 cm x 4.5 cm rectangular cake.

Tips

1 The almond shortbread pastry is supposed to be crispy, so as to add a texture substantially different from sponge cake and the mousse. Shortbread dough usually calls for icing sugar because it melts faster in butter than granulated sugar. The dough will end up finer in texture this way. On the other hand, granulated sugar does not melt in the mixing process. It tends to melt when you roll the dough out so that the dough gets torn easily and the melted sugar sticks to the counter and rolling pin. The problem gets even more serious when the weather is hot. To solve the problem, you may sprinkle flour repeatedly on the counter before each rolling. But still, the pastry would end up having a thick crust after baked, which is not desirable.

For the shortbread pastry to be crispy, the key lies in preventing the gluten from forming in the kneading and mixing stage. Try to keep flour away from water. Or mix in the flour right before you bake it to minimize the time the two get in contact.

2 Exotic jelly filling: If you can't get the purees, just fresh fruit juices instead.

3 Fromage frais Onctueux is preferred. Fromage frais is a French fresh white cheese and it comes in two varieties: fat-free and with 40% solid fat. To make the cake velvety smooth, please pick the latter.

4 Adding a little sugar to the egg yolk first helps temper it and prevent the egg yolk from being cooked and turned into lumps when hot syrup is poured into it.

5 The syrup is heated up to 118°C before poured into raw egg yolk and beaten over high speed. This temperature is hot enough to kill most salmonella bacteria in the egg yolk while bringing out its authentic taste.

6 When you stir in the whipped cream, do not over-stir it. Use a plastic spatula instead of any electric mixer.

草莓忌廉蛋糕
Strawberry Cream Cake

雪白可愛如小女孩的紗裙,是所有初學者都想做的蛋糕;看似簡單,卻考理論、考工夫。製作海綿蛋糕可用分蛋法和全蛋法:全蛋,糕綿密細緻;分蛋,則鬆軟有彈性,也較易掌握;可視乎自己口味和技術來製作所需蛋糕體。

製作次序	1. 全蛋法海綿蛋糕
	2. 香桃酒糖水
	3. 裝組

全蛋法海綿蛋糕
Sponge Cake

材料

全蛋	180 克
砂糖	105 克
低筋麵粉	110 克
無鹽牛油(室溫放軟)	30 克
牛奶	30 克

** 可製成一個 15 厘米 x 7.5 厘米蛋糕

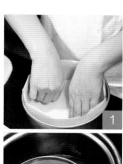

做法

1. 於金屬模底和邊墊上牛油紙;麵粉過篩三次。(看圖 1)

2. 牛油與牛奶一同煮溶,並將溫度保持於 50℃ 左右。(看圖 2)

3. 蛋與砂糖坐於 60℃ 熱水上,打發至溫度約 36℃。❶ (看圖 3)

4. 再以高速打發 6-7 分鐘至濃稠,再轉低速繼續打發 2-3 分鐘以穩定泡沫。❷ (看圖 4)

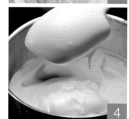

5. 用膠刮以切入的手法把麵粉混合蛋糊。

6. 先舀小部分麵糊與(2)混合,再放入其餘麵糊拌勻。❸

7. 倒進金屬模內,用已預熱 170℃ 的焗爐焗約 20-25 分鐘。(看圖 5)

香桃酒糖水
Peach Syrup

材料

香桃酒	15 克
砂糖	30 克
熱水	100 克

做法

將砂糖放入熱水內，攪拌至溶化，加入香桃酒，拌勻。

裝飾

草莓（餡料用）（切片）	適量
草莓（裝飾用）	適量
覆盆子	適量

打起甜忌廉 100 克、淡忌廉 400 克（混合後一同打發）❹

裝組

1. 把海棉蛋糕橫切成三層。（看圖 6）❺❻

2. 部分忌廉打至挺身作擠花用，另一部分則只需打至八成起作「扲餅」用。

3. 每層蛋糕先噴上香桃酒糖水，再抹一層忌廉，上面放草莓片。（看圖 7-13）

** 可做一個 15 厘米 x 10 厘米蛋糕

貼士

❶ 要製作一個優質的全蛋法蛋糕，蛋加糖後需加熱至 36℃，如此才不受天氣冷暖影響，任何情況下都能保持水準。糖在蛋內溶化，較容易打至濃稠，加入麵粉時便容易混合。

❷ 打發全蛋時，初段要使用高速，泡沫會越來越多，體積不斷增大，一直打約 6-7 分鐘，看到泡沫感覺有彈性，又輕又軟，但此時泡沫大小不一，亦未穩定，這時將打蛋器調至慢速，大約再打發約 2-3 分鐘，直至泡沫細緻光亮。

❸ 當麵糊混合牛油後，蛋泡就會很易消散，所以切勿過度攪拌。

❹ 淡忌廉香滑但奶味過重，口味過膩，而且做餅面裝飾不夠穩定，容易分離和溶掉。甜忌廉乳脂含量較少，口感輕盈，可是奶味不足，而且過甜，但打發後結構穩定，故把兩種忌廉混合使用為佳。以淡忌廉所佔比例較多，約 7-8 份淡忌廉配 2-3 份甜忌廉一起打發最為理想。做餅面裝飾要注意，「抆餅」的忌廉只要打至八成就可以，因為抆在餅上的忌廉會被來回推動，這動作會令忌廉老化，表面就不光滑。如果想擠花，就把用剩的忌廉用打蛋器再稍加打發至挺身。

❺ 把蛋糕分層最好使用轉台。看準要分開層數的厚度，用長牙刀輕輕剠開，以刀口為切口點，長牙刀向前後移動，另一隻手轉動轉台，切開蛋糕。

❻ 握抆刀姿勢必須正確，緊握手柄較下位置，食指緊按刀刃，這樣才能靈活使用抆刀。(看圖 14)

Strawberry Cream Cake

Production workflow

1. Make the sponge cake.
2. Make the peach syrup.
3. Assemble and garnish.

Sponge Cake

Ingredients

180 g eggs
105 g sugar
110 g cake flour
30 g unsalted butter (at room temperature)
30 g milk
** Makes a 15 cm x 7.5 cm sponge cake

Method

1. Line the bottom and the vertical sides of the cake tin with parchment paper. Sift the flour three times. (see picture 1)
2. Heat the milk and butter up to 50˚C. Keep them around that temperature. (see picture 2)
3. Leave the eggs and sugar in a water bath at 60˚C. Mix well while heating it up to 36˚C. (see picture 3)❶
4. Beat the eggs over high speed for 6 to 7 minutes until thick and fluffy. Then turn to low speed and beat for 2 to 3 minutes to stabilize the foam. (see picture 4)❷
5. Fold the sieved flour into the egg mixture.
6. Scoop out a small ladle of the egg mixture. Pour into the hot milk and butter mixture from step 2 to temper it first. Mix well. Pour all of the mixture back into the egg mixture to mix well.❸
7. Pour into the cake tin. Bake in a preheated oven at 170˚C for 20 to 25 minutes. (see picture 5)

Peach Syrup

Ingredients

15 g peach liqueur

30 g sugar
100 g boiling water

Method

Add sugar to water. Stir until it dissolves. Add peach liqueur. Stir well.

Garnish

sliced strawberries (to be sandwiched between the sponge layers)

strawberries

raspberries

100 g non-dairy whipped topping❹

400 g whipping cream

Assembly

1. Slice the sponge cake horizontally into three layers. (see picture 6)❺❻

2. Mix together the non-dairy whipped topping and whipping cream. Beat until soft peaks form. Set aside part of it for spreading. Beat the remaining mixture until stiff for piping.

3. Spray peach syrup on each slice of sponge. Spread whipped cream mixture over each of them. Arrange sliced strawberries over the whipped cream. Stack the three slices of sponge on top of each other. Spread whipped cream mixture all over the cake. Garnish with strawberries and raspberries. (see pictures 7-13)

** Makes one 15-cm round cake about 10 cm tall.

Tips

❶ To make a sponge cake without separating the egg yolks and whites and without adding any rising agent, you have to heat the egg and sugar mixture up to 36°C first before beating. This is the optimal temperature and you don't need to worry about the room temperature this way. The sugar tends to melt into the eggs so that they thicken more easily. The flour and eggs also tend to combine more easily.

❷ Beat the eggs over high speed in the beginning. The egg mixture will turn foamy and expand in volume. After beating at high speed for 6 to 7 minutes, the foam should look resilient, light and fluffy. But the bubbles still vary in sizes and tend to burst easily. It is time to turn the mixer to low speed and beat for 2 or 3 more minutes. You should end up with silky smooth bubbles with a glossy sheen.

❸ After adding butter to the egg mixture, the bubbles tend to break easily. Do not over-stir and do not stir too vigorously.

❹ Whipping cream is smooth in texture, but maybe too heavy for some. As a frosting on the cake, whipped cream itself is not stable enough. It tends to break down and turns watery after a while. Non-dairy whipped topping contains no milk fat at all and it is light and highly stable. However, it lacks the richness and milky flavour of whipping cream. It is also a bit too sweet on its own. Thus, for the best result, I suggest mixing 7 to 8 parts of whipping cream with 2 to 3 parts of non-dairy whipped topping. Finally, the whipped cream mixture should be beaten to different stages for different purposes. For cream to be spread over the cake, beat until soft peaks form. As you spread and smear the cream, the cream will further be stiffened. If it is stiff to begin with, the cream might become dry and over-beaten after you push it around a few times. For cream to be piped, beat until stiff.

❺ To cut the cake into layers, it is advisable to use a cake turntable or a lazy Susan. Measure how thick you want each layer to be. Mark the cut with a serrated knife. Insert the knife horizontally on the mark. Run the blade back and forth while turning the turntable slowly with the other hand. You'd end up with a clean cut this way.

❻ You should also pay attention to the way you hold a palette knife. Hold the handle close to where it connects with the blade and press the blade with your index finger. You can then control the motion more articulately. (see picture 14)

黑森林蛋卷
Black Forest Log

朱古力海綿蛋糕、黑朱古力忌廉和酒浸罐頭車厘子組成的省時超懶人蛋糕，臨急也不用抱佛腳了。

製作次序

1. 酸車厘子餡心
2. 朱古力海綿蛋糕
3. 朱古力忌廉
4. 裝組

酸車厘子餡心
Sour Cherry Filling

材料

Aldia 罐裝櫻桃酒浸車厘子餡料 ❶	250 克
水	40 克
砂糖	10 克
魚膠片（用凍水浸軟）	4 克

做法

1. 水和砂糖煮滾，放入已浸軟的魚膠片，攪拌至完全溶化。

2. 加入罐裝櫻桃酒浸車厘子餡料內，拌勻，倒入模中冷凍至結冰。（看圖 1-2）

朱古力海綿蛋糕 ❷
Chocolate Sponge Cake

材料

蛋黃	80 克	砂糖 (B)	60 克
全蛋	180 克	低筋麵粉	50 克
砂糖 (A)	150 克	可可粉	40 克
蛋白	120 克	溶無鹽牛油（暖和）	30 克

** 預備 30 厘米 x 40 厘米長方型焗盤一個
   或 30 厘米 x 20 厘米長方型焗盤兩個

做法

1. 麵粉、可可粉混合過篩。

2. 全蛋、蛋黃和砂糖 (A) 打至淡黃色。（看圖 3）

3. 蛋白和砂糖 (B) 打發成柔韌細緻的蛋白霜。（看圖 4）

4. 將 (2) 加入 (3)，以切拌方式輕輕拌勻。

5. 將粉類輕且快地切拌入 (4)**❸**，盡量避免蛋白消泡，然後拌入暖牛油溶液。（看圖 5）

6. 把約一碗混合料舀進盛牛油的容器內，拌勻，再倒回 (5) 內，輕輕拌勻。（看圖 6）

7. 倒入墊上矽膠蓆或不黏布的烤盤內，抹平，以 180℃ 焗 10-12 分鐘。（看圖 7-8）

朱古力忌廉 **❹**

Chocolate Chantilly Cream

材料

黑朱古力（50% 谷咕含量以上）	200 克
打起淡忌廉	500 克
冧酒	適量

做法

1. 黑朱古力隔熱水坐溶或以微波爐煮溶，直至朱古力溫度約為 55℃ - 60℃。（看圖 9）

2. 將已溶化的朱古力快速拌入已打起淡忌廉，最後拌入冧酒。（看圖 10-11）

裝組

1. 準備兩個 20 厘米 x 9 厘米 x 6 厘米半圓餅模，用保鮮紙包好模具底部，依模具大小切割一份朱古力海綿蛋糕放在模底。

2. 噴少許櫻桃酒糖水使蛋糕濕潤。

3. 擠入半份黑朱古力忌廉，放上酸車厘子餡心。

4. 再填滿黑朱古力忌廉。

5. 最後鋪上一片朱古力海綿蛋糕，冷藏凝固後脫模，用朱古力碎和鮮車厘子作裝飾。（看圖 12-17）

** 可做兩卷 20 厘米 x 9 厘米 x 6 厘米半圓長條蛋糕

貼士

❶ 罐裝櫻桃酒浸車厘子餡料，在烘焙食材店有散裝出售，它是一種附有醬汁的櫻桃酒酸車厘子餡料，可用作蛋糕、派餅和餅面，方便好用。

❷ 此食譜的朱古力海綿蛋糕以分蛋法製作，蛋要室溫。蛋白要清澈乾淨，不能混入蛋黃；打發的器皿不可沾有水和油分，否則蛋白不能打發起泡。

❸ 加入粉類後要輕且快地拌勻，因可可粉油很重，很易令蛋白消泡，故此動作要快速；可將小部份麵糊先混入暖牛油中，攪拌均勻才與全部麵糊混合，牛油就不易沉底。

❹ 朱古力要隔着熱水坐溶，間中攪拌，直至朱古力溫度達 55-60℃，立即混入打發淡忌廉內快速攪拌。如果朱古力溫度太低，混入冷的忌廉時就會結成硬塊，忌廉就不幼滑了。

Black Forest Log

Production workflow

1. Make the sour cherry filling.
2. Make the chocolate sponge cake
3. Make the chocolate chantilly cream
4. Assemble the cake.

Sour Cherry Filling

Ingredients

250 g "Aldia" brand cherry pie filling in Kirsch **❶**

40 g water

10 g sugar

4 g gelatine leaf (soaked in cold water until soft; drained)

Method

1. Bring water and sugar to the boil. Add soaked gelatine leaf and stir until gelatine dissolves.
2. Stir in cherry pie filling. Mix well. Pour onto the mould and freeze it.

Chocolate Sponge Cake **❷**

Ingredients

80 g egg yolk

180 g eggs

150 g sugar (A)

120 g egg whites

60 g sugar (B)

50 g cake flour

40 g cocoa powder

30 g melted unsalted butter

** Prepare one baking tray measuring 30 cm x 40 cm; or two baking trays measuring 30 cm x 20 cm.

Method

1. Sift flour and cocoa powder into a mixing bowl.
2. In another bowl, beat eggs, egg yolks and sugar (A) together until pale. (see picture 3)
3. In a separate bowl, beat egg whites and sugar (B) until stiff. (see picture 4)
4. Pour the egg yolk mixture from step 2 into the egg whites from step 3. Fold gently until well mixed.
5. Fold in dry ingredients into the mixture from step 4.**❸**Try not to destroy the air bubbles in the stiff egg whites. Then fold in the warm butter. (see picture 5)
6. Scoop 1 ladle of batter into the butter container. Mix well and pour it back in the batter. Fold gently. (see picture 6)
7. Pour the batter into a baking tray lined with silicone mat or non-stick baking lining. Smooth the top out. Bake in a preheated oven at 180°C for 10 to 12 minutes. Leave it to cool. (see pictures 7-8)

Chocolate Chantilly Cream **❹**

Ingredients

200 g dark chocolate (over 50% cocoa)

500 g whipped cream

rum

Method

1. Melt the chocolate over a hot water bath or in microwave up to 55°C-60°C. (see picture 9)
2. Quickly stir in whipped cream. Stir in rum. (see pictures 10-11)

Assembly

1. Prepare two log moulds, each measuring 20 cm x 9 cm x 6 cm. Line the rounded bottom of your mould with cling film. Then trim a slice of chocolate sponge to fit in the

rounded bottom of the mould. Press the chocolate sponge into the mould.

2. Spray on some Kirsch syrup to make the cake moist.

3. Pipe half of the chocolate chantilly cream over the sponge. Arrange the frozen sour cherry filling over the cream.

4. Fill the inside of the sponge with remaining chocolate chantilly cream.

5. Top it with a slice of chocolate sponge. Refrigerate until set. Unmould. Garnish with chocolate shavings and fresh cherries. Serve. (see pictures 12-17)

** Makes 2 log cakes

Tips

❶ Canned cherry pie filling in Kirsch is available from bakery supply stores. As its name suggests, it is a readymade pie filling made with sour cherry steeped in cherry brandy. It is commonly used in cakes, pies and glazing.

❷ The chocolate sponge in this recipe calls for separated egg whites and yolks. Bring the eggs to room temperature first before separating. The egg whites have to be completely free from any egg yolk. Make sure the container and tools that you use to beat the egg whites are free from any grease or water. Otherwise, the egg whites won't stand.

❸ After adding dry ingredients, fold quickly but gently because the cocoa oil is heavy and is likely to drive the air out of the stiffly beaten egg whites. Pour a little batter into the warm butter and mix well first, before pouring it back into the batter. The butter is less likely to separate from the batter this way.

❹ To melt the chocolate, place it in a bowl over a pot of simmering water. Stir occasionally until the chocolate reaches 55°C–60°C and remove from heat. Then pour in the whipped cream and mix well swiftly. If the chocolate is not warm enough, it sets right away when it hits the cold whipped cream and makes the cream lumpy.

杏桃圓餅
Apricot Dome

杏桃圓餅巧妙地把杏桃與其種子杏仁再次結合為一體，既有濃濃的杏仁堅果香，同時洋溢杏桃的清新氣息。

<table>
<tr><td rowspan="3">製作次序</td><td>1. 杏桃餡心</td><td>4. 杏桃果茸鏡面</td></tr>
<tr><td>2. 杏仁慕絲</td><td>5. 益壽糖裝飾</td></tr>
<tr><td>3. 開心果蛋白餅</td><td>6. 裝組</td></tr>
</table>

杏桃餡心
Apricot Coulis

材料

冷凍杏桃果茸	150 克
檸檬汁	3 克
砂糖	40 克
粟粉 ❶	7 克
魚膠片（用凍水浸軟）	3 克
水	14 克
冷凍開邊杏桃 ❷	100 克

做法

1. 將冷凍開邊杏桃用牛油和砂糖（份量以外，各約 20 克）炒至軟，再剪成塊狀。（看圖 1-2）

2. 杏桃果茸、檸檬汁和水煮滾。

3. 粟粉、砂糖拌勻，拌入（2）煮至濃稠。

4. 加入已浸軟魚膠片，拌勻，下開邊杏桃，拌勻，再倒入直徑約 13 厘米、深 3 厘米的容器中，冷凍至結冰。（看圖 3）

杏仁慕絲
Almond Amaretto Cream

材料

牛奶	80 克
淡忌廉	80 克
雲呢拿豆莢	1 枝
（用小刀刮出雲呢拿籽）	
杏仁膏 (50%)	80 克
蛋黃	40 克
砂糖	10 克
粟粉	4 克
魚膠片（用凍水浸軟）	7 克
意大利苦杏酒 ❸	20 克
打起淡忌廉	200 克

做法

1. 鮮奶、淡忌廉混合，煮滾。加入雲呢拿籽及杏仁膏同煮，直至杏仁膏完全融化。

2. 蛋黃、砂糖和粟粉拌勻，倒入 (1)。

3. 過篩後回鍋煮滾，期間不斷攪拌，加入已浸發的魚膠片。

4. 冷卻後拌入意大利苦杏酒，拌入已打起淡忌廉。

開心果蛋白餅 ❹
Pistachio Dacquoise

材料

蛋白	106 克
砂糖	84 克
杏仁粉	84 克
低筋麵粉	20 克
糖霜	40 克
開心果醬	25 克

做法

1. 焗爐預熱約 160℃；麵粉過篩；焗盤墊上矽膠墊或不黏布，放上直徑 20 厘米的圓環餅模，餅模內噴上少許食用油噴劑。（看圖 4）

2. 糖霜、麵粉和杏仁粉拌勻，可用手輕輕搓揉均勻。（看圖 5）

3. 蛋白和砂糖打至柔韌有力。（看圖 6）

4. 把杏仁糖粉輕輕拌入蛋白霜內，然後拌入開心果醬。（看圖 7）

5. 把麵糊舀入擠花袋，擠入圓環餅模內，灑上適量糖霜，放入已預熱 160℃ 的焗爐內焗約 25 分鐘。（看圖 8-10）

杏桃果茸鏡面
Apricot Glazing

材料

鏡面果膠 ❺	600 克
杏桃果茸	80 克

做法

1. 所有材料隔熱水坐暖或用微波爐煮熱。

2. 用電動棒狀攪拌器慢速打滑，再放涼至 50℃。

益壽糖裝飾
Isomalt Decoration

材料

益壽糖	80 克

做法

1. 益壽糖放在兩片矽膠蓆中，以 160℃ 度焗約 20 分鐘或糖溶化。（看圖 11-12）

2. 待糖完全冷卻才從矽膠蓆中取出。

裝飾

燒杏桃

裝組

1. 杏仁慕絲倒入直徑 15 厘米，高 9 厘米的圓拱形容器中，至 1/2 滿時放上杏桃餡心，再倒進剩餘杏仁慕絲至滿，放入冰格冷凍至結冰。（看圖 13-15）

2. 脫模後，淋上杏桃果茸鏡面，放在開心果蛋白餅上，最後放上燒杏桃和益壽糖裝飾。（看圖 16-20）

貼士

1. 粟粉在中式菜餚常作打獻之用，用水調勻再混入菜餚或汁水內令其濃稠；如直接將粟粉加入熱水內會結塊。這裏的做法是將粟粉混合少量砂糖，放入熱液體再煮，可避免結塊。

2. 冷藏杏桃果茸和冷藏開邊杏桃在 DIY 店有售，亦可用新鮮或罐頭杏桃代替。

3. 杏仁慕絲添加了意大利苦杏酒後能突出杏仁香氣，如買不到可省去。

4. Dacquoise 是沒有蛋黃的蛋白餅，要做到外脆內煙韌，攪打的蛋白需要先冷藏，讓蛋白變得較稀。糖量必須超過蛋白的一半，才能打發成柔韌有力的蛋白霜。

5. 鏡面果膠宜購買需加熱處理的。這類果膠通常有原味透明、杏桃味（黃色）和草莓（紅色）三種，皆可加水或其他液體煮熱作裝飾用，例如塗於生果表面，使其閃閃生輝。

Apricot Dome

Production workflow

1. Make the apricot coulis.
2. Make the almond amaretto cream.
3. Make the pistachio dacquoise.
4. Make the apricot glazing.
5. Make the Isomalt decoration.
6. Assemble the cake.

Apricot Coulis

Ingredients

150 g frozen apricot puree

3 g lemon juice

40 g sugar

7 g cornflour ❶

3 g gelatine leaf (soaked in water until soft; drained)

14 g water

100 g frozen apricot halves ❷

Method

1. Thaw the frozen apricot halves. Sauté them in about 20 g of butter and 20 g of sugar (not included in the quantity above). Cut into wedges. (see pictures 1-2)
2. Bring apricot puree, lemon juice and water to the boil.
3. Mix cornflour and sugar well. Add cornflour-sugar mixture to the apricot mixture from step 2. Cook until thick.
4. Add the gelatine leaf to the apricot mixture from step 3. Stir until it dissolves. Add apricot wedges. Mix well. Pour into a 13-cm round and 3-cm depth container. Keep in the freezer until frozen. (see picture 3)

Almond Amaretto Cream

Ingredients

80 g milk

80 g whipping cream

1 vanilla pod (split opened; seeds scraped out with a knife)

80 g almond paste (50 %)

40 g egg yolk

10 g sugar

4 g cornflour

7 g gelatine leaf (soaked in water until soft; drained)

20 g Amaretto ❸

200 g whipped cream

Method

1. Boil the milk and whipping cream. Put in almond paste and vanilla seeds. Cook until the almond paste melts.
2. Whisk the egg yolk, sugar and cornflour. Pour the boiling milk mixture from step 1 into the egg yolk mixture.
3. Sift the resulting mixture and continue to cook until it boils again. Add the gelatine leaf and cook until it dissolves.
4. Leave the mixture to cool. Add Amaretto. Fold in whipped cream.

Pistachio Dacquoise ❹

Ingredients

106 g egg whites

84 g sugar

84 g ground almonds

20 g cake flour

40 g Icing sugar

25 g pistachio paste

Method

1. Preheat the oven at 160°C. Sift flour into a mixing bowl. Line a baking tray with non-stick baking liner or silicone mat. Put a 20 cm ring on the tray. Grease the mould with pan release spray. (see picture 4)
2. Mix icing sugar, flour and ground almonds together. You may rub them slightly with your hand. (see picture 5)

3. Whip egg whites and sugar until stiff. (see picture 6)
4. Fold in ground almond-flour mixture from step 2. Add pistachio paste. (see picture 7)
5. Pipe into the ring. Sprinkle with icing sugar. Bake at 160°C at 25 minutes. Leave it to cool. (see pictures 8-10)

Apricot Glazing

Ingredients

600 g neutral glazing ❺
80 g apricot puree

Method

1. Warm all ingredients together over a pot of simmering water or in a microwave.
2. Blend over low speed until smooth. Leave it to cool to about 50°C.

Isomalt Decoration

Ingredients

80 g Isomalt

Method

1. Put the Isomalt between two silicone baking mats. Bake in an oven at 160°C for about 20 minutes or until it melts. (see pictures 11-12)
2. Wait over low speed until the Isomalt is cool. Remove the Isomalt from the silicone mats for decoration.

Garnish

caramelized apricot

Assembly

1. Prepare a dome mould 15 cm in diameter and 9 cm deep. Fill the mould with the almond amaretto cream up to 1/2 of its depth. Put in the frozen apricot coulis. Fill the mould up. Keep it in the freezer until frozen. (see pictures 13-15)
2. Unmould the frozen almond amaretto cream. Pour apricot glazing over the dome. Put the dome-shaped glazed mousse cake over the pistachio dacquoise. Garnish with caramelized apricot wedges and Isomalt decorations. Serve. (see pictures 16-20)

Tips

❶ Cornflour is commonly used as a thickening agent in Chinese cooking. But adding cornflour directly to boiling liquid would make it lumpy. Thus, usually it is thinned out in some water first, before stirred into a food, sauce or soup. In this recipe, the cornflour is mixed with sugar first before added to hot liquid, so as to prevent lumps from forming.

❷ Frozen apricot puree and apricot halves are available from stores specializing in DIY baking needs. You may also use fresh or canned apricot instead.

❸ Adding amaretto (an Italian almond liqueur) to the almond cream helps accentuate the aroma. If you can't get it, you may skip it.

❹ Dacquoise is a meringue. To make it crispy on the outside and chewy on the inside, you should refrigerate the egg whites to make them thinner in consistency before beating them. The sugar should weigh more than half the weight of egg whites for fluffy and stiff meringue.

❺ Try to get neutral glazing that needs to be heated up before application. Glazing is available in three flavours and colours: neutral (transparent); apricot (yellow) and strawberry (red). You may add water or juice to them and then heat it up and apply on cakes as decoration. It adds a gem-like sheen to fruits, for instance.

柚子撻
Yuzu Tart

酸又滑的日本柚子醬配上甜且稠的意大利蛋白霜，有如陰陽，相生相尅，缺一不可。有別於一般檸檬撻的製法，這種餡料口感順滑，牛油份量也較少，也不需再次入爐。

<table>
<tr><td rowspan="3">製作次序</td><td>1. 杏仁脆餅底</td><td>4. 裝組</td></tr>
<tr><td>2. 日本柚子醬</td><td></td></tr>
<tr><td>3. 意大利蛋白霜</td><td></td></tr>
</table>

杏仁脆餅底 ❶
Almond Pie Crust

材料

牛油	50 克
糖霜	40 克
全蛋	20 克
低筋麵粉	50 克
高筋麵粉	50 克
杏仁粉	20 克

做法

1. 牛油和糖霜用刮刀拌勻，直至砂糖溶化。

2. 逐少加入雞蛋，繼續拌勻。

3. 輕輕拌入已篩麵粉及杏仁粉，注意不要過分搓揉，否則餅底會過硬。

4. 將麵糰放入雪櫃最少一小時或過夜。從雪櫃取出後立即擀成撻皮，圍在撻模內，用义刺上小孔，雪硬。（看圖 1-3）

5. 錫紙墊在撻皮上，放一些已烘熱的金屬珠或豆類壓著撻皮，放進已預熱 180℃ 的焗爐內，焗約 12-15 分鐘至金黃色。（看圖 4）

6. 焗至差不多熟時取出，塗一層蛋黃，再放入焗爐多焗約 3 分鐘，放涼備用。❷（看圖 5）

日本柚子醬 ❸
Yuzu Curd

材料

冷凍日本柚子果汁	100 克
或	
新鮮日本柚子汁	
砂糖（A）	60 克
無鹽牛油（室溫放軟）	90 克
全蛋	120 克
砂糖（B）	50 克
冷凍日本柚子皮	10 克
或	
新鮮日本柚子皮茸	
魚膠片（凍水內浸軟）	3 克

做法

1. 柚子果汁、砂糖（A）和牛油煮滾。

2. 雞蛋與砂糖（B）拌勻，邊拌邊慢慢注入（1）。

3. 將（2）過篩，回鍋煮滾，期間不斷攪拌，過篩，離鍋後加入日本柚子皮。

4. 加入已浸軟的魚膠片，拌勻。

意大利蛋白霜 ❹
Italian Meringue

材料

水	30 克
砂糖（A）	180 克
蛋白	80 克
砂糖（B）	30 克

做法

1. 水和砂糖（A）一同煮滾至 120℃。

2. 蛋白及砂糖（B）打起。

3. 將煮滾的糖水緩緩加入蛋白糊裏，並繼續攪打至光亮而柔韌有力和溫度降低，放雪櫃雪凍。（看圖 6-8）

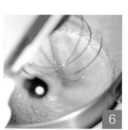

裝組

1. 用隔篩將撻皮磨滑。（看圖 9）

2. 日本柚子醬倒入撻殼，放入雪櫃雪至凝固。
（看圖 10）

3. 將打好的意大利蛋白霜用擠花袋擠在撻面
上，再用火槍燒至蛋白霜發脹焦黃。（看圖
11-13）

** 可做一個 18 厘米撻

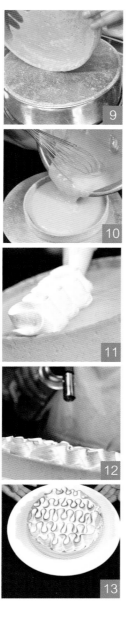

貼士

❶ 杏仁脆餅麵糰需預先製作，放入
雪櫃最少一小時或過夜。

❷ 蛋黃的油分會形成一層防水的保
護膜，令撻皮不易因受潮而軟化。
如製作超過一個撻皮，請逐份從
雪櫃取出撻皮及造型，尤其室溫
高時更要快速製作，才不會令麵
糰過軟；麵糰變軟時須把麵糰再
放雪櫃雪硬才使用；不要重複的
擀捲或摺疊麵糰，會使成品過硬，
不鬆脆。

❸ 日本柚子醬的酸度可自由調節，
只要增加或減少柚子汁的份量即
可。

❹ 小於一個蛋白的意大利蛋白霜很
難操作，建議最少做兩個蛋白。
剩餘的蛋白霜可儲存於冰格一個
月。食譜中的糖量已經很少，不
能再減，否則蛋白霜會口感粗糙、
沒有光澤和容易分離。

Yuzu Tart

Production workflow

1. Make the almond pie crust.
2. Make the Yuzu curd.
3. Make the Italian meringue.
4. Assemble.

Almond Pie Crust ❶

Ingredients

50 g butter
40 g icing sugar
20 g eggs
50 g cake flour
50 g bread flour
20 g ground almonds

Method

1. Stir butter and icing sugar with a plastic spatula until all icing sugar is moistened by the butter.
2. Pour in the egg a little at a time. Stir well after each addition.
3. Sift in flour and ground almonds. Stir gently. Do not over-stir. Or else the pie curst will be hard, instead of crispy.
4. Leave the dough in the fridge for at least one hour (preferably overnight). Roll the dough out into a thin sheet. Press it into the tart mould. Prick holes all over with a fork. Refrigerate until firm. (see pictures 1-3)
5. Line the pie crust with a sheet of aluminium foil. Then put pre-heated baking beans over the aluminium foil. Bake in a preheated oven at 180°C for 12 to 15 minutes until golden. (see picture 4)
6. Brush on a thin layer of egg yolk on the pie crust. Bake in the oven for 3 more minutes. Set aside to let cool. (see picture 5) ❷

Yuzu Curd ❸

Ingredients

100 g Yuzu juice (frozen or fresh)
60 g sugar (A)
90 g unsalted butter (at room temperature)
120 g eggs
50 g sugar (B)
10 g grated Yuzu zest (frozen or fresh)
3 g gelatine leaf (soaked in cold water until soft; drained)

Method

1. Cook Yuzu juice, sugar (A) and butter until it boils.
2. In a bowl, mix eggs and sugar (B) together. Pour the hot Yuzu juice mixture from step 1 into the egg mixture while stirring continuously.
3. Pass the resulting mixture through a sieve. Pour back into a pot and bring to the boil while stirring constantly. Remove from heat. Sift again. Add Yuzu zest.
4. Put in the gelatine leaf. Stir until it dissolves.

Italian Meringue ❹

Ingredients

30 g water
180 g sugar (A)
80 g egg whites
30 g sugar (B)

Method

1. Cook water and sugar (A) until it reaches 120°C.
2. In a mixing bowl, beat egg whites and sugar (B) until stiff.

3. Pour the hot syrup from step 1 slowly into egg whites from step 2. Keep on beating until satiny and ribbon-like. Refrigerate. (see pictures 6-8)

Assembly

1. File off the irregular edges of the pie crust with a sieve or a grater. (see picture 9)
2. Pour the Yuzu curd into the pie crust. Refrigerate until set. (see picture 10)
3. Transfer the Italian meringue into a piping bag. Pipe the meringue over the Yuzu curd. Brown the meringue with a propane torch. Serve. (see pictures 11-13)

**Makes one 18-cm round tart

Tips

❶ The almond pie crust dough needs time to rest. Leave it in the fridge for at least 1 hour, or preferably overnight, before use.

❷ The grease in the egg yolk forms an impervious coating on the pie crust, so that it won't turn soggy after getting in touch with the filling. If you're making more than one pie at a time, take only enough dough out of the fridge to make one pie first. Keep the remaining dough in the fridge. Wait till you're done with the first pie crust before taking more dough out of the fridge. Make sure you work swiftly with the dough, especially when the weather is hot. If you find the dough a bit too soft to keep its shape, refrigerate again until firm. Do not over-fold or roll the dough, or else the crust will be hard instead of crispy.

❸ You may adjust the sourness of the Yuzu curd according to your own taste. Just adjust the amount of Yuzu juice to your liking.

❹ It's quite difficult to work with just one egg white when you're making meringue. Thus, I suggest using two egg whites. The leftover meringue lasts well in the freezer for 1 month. The amount of sugar in the meringue is already the bare minimum, meaning you can't cut down on it further. Otherwise, the meringue will end up in coarse and dull texture. It also tends to separate easily making your meringue wet and sticky.

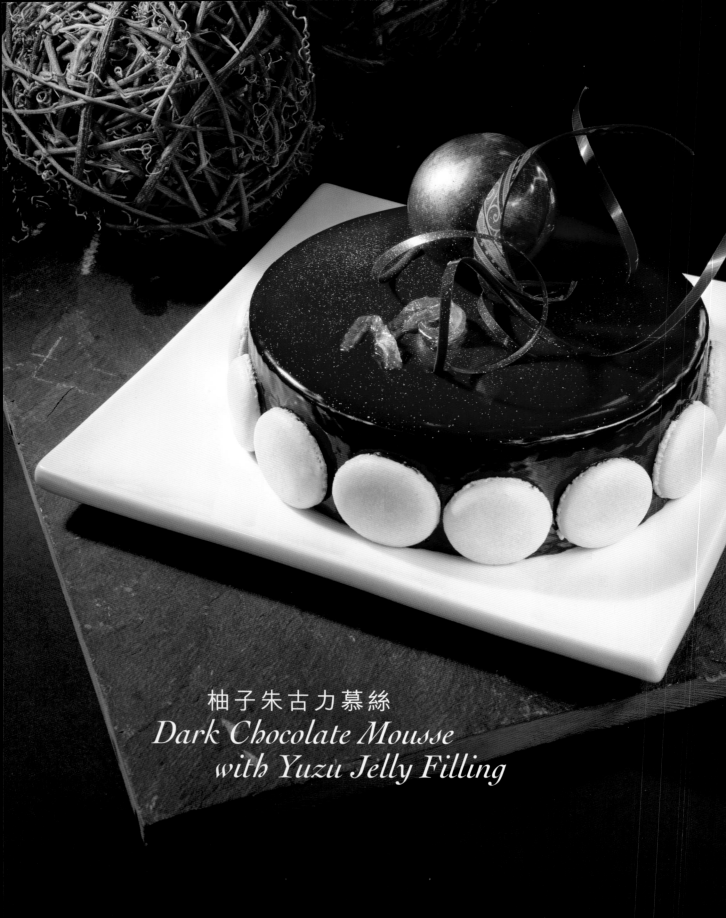

柚子朱古力慕絲
*Dark Chocolate Mousse
with Yuzu Jelly Filling*

濃香與馥郁，激情與爽朗。黑朱古力與柚子，不是偶然的拼湊，而是一段異國情緣。

製作次序	1. 柚子餡心	4. 黑朱古力慕絲	7. 裝組
	2. 朱古力手指餅	5. 黑朱古力淋面	
	3. 黑朱古力脆脆餅底	6. 柚子糖漿	

柚子餡心
Yuzu Filling

材料

冷凍柚子果汁 ❶	40 克
水	35 克
砂糖	50 克
魚膠片 (用凍水浸軟)	4 克
冷凍柚子皮	20 克

做法

1. 水和砂糖煮滾，離火加入柚子果汁、冷凍柚子皮和已浸軟魚膠片，拌勻。（看圖 1-2）

2. 倒入 1 厘米的圓模內，放入冰格冷凍至結冰，待用。（看圖 3）

** 可做 1 片 1 厘米 x 15 厘米餡心

朱古力手指餅
Chocolate Ladyfinger

材料

高筋麵粉	50 克	蛋白	100 克
可可粉	15 克	砂糖 (B)	30 克
蛋黃	60 克	粟粉	40 克
砂糖 (A)	30 克	砂糖 (C)	30 克

做法

1. 請參考第 42 頁的咖啡慕絲內手指餅的做法（1）至（4）。

2. 麵糊舀進擠花袋內，然後擠 1 個直徑 18 厘米的圓餅在已墊不黏布或矽膠蓆的焗盤上，篩上糖霜（份量外）。（看圖 4-6）

3. 放入已預熱 170℃ 的焗爐內，焗約 15 分鐘。

黑朱古力脆脆餅底
Chocolate Crispy Base

材料

薄脆片	80 克
黑朱古力（50% 谷咕含量以上）	50 克
榛子醬	50 克

做法

1. 將黑朱古力坐熱水或在微波爐溶化（約 50℃）❷。（看圖 7-8）

2. 下榛子醬，拌勻；加入薄脆片，拌勻。（看圖 9）

3. 壓入直徑 15 厘米的餅模內，雪硬待用。（看圖 10）

黑朱古力慕絲
Dark Chocolate Mousse

材料

黑朱古力（50% 谷咕含量以上）(切碎)	160 克
淡忌廉	100 克
牛奶	100 克
砂糖	20 克
蛋黃	40 克
打起淡忌廉	250 克
魚膠片 (用凍水浸軟)	5 克

做法

1. 調勻蛋黃及砂糖。

2. 淡忌廉、牛奶煮滾。

3. 將（2）倒入（1）內拌勻，過篩後加熱至 85℃。（看圖 11）

4. 將（3）直接篩在朱古力上，拌勻，待朱古力溶化後，加入已浸軟的魚膠片，拌勻，待完全涼透。（看圖 12-13）

5. 分數次拌入打起淡忌廉。（看圖 14-15）

黑朱古力淋面
Dark Chocolate Glaze

材料

水	150 克
砂糖	120 克
淡忌廉	120 克
魚膠片（用凍水浸軟）	10 克
可可粉	60 克

做法

1. 水、忌廉和砂糖煮滾。加入可可粉,繼續煮至滾。
   (看圖 16-17)

2. 下已浸軟的魚膠片,拌勻,過篩,放涼至約
   30℃,待用。(看圖 18-19)

柚子糖漿
Yuzu Syrup

材料

冷凍柚子汁	20 克
砂糖	60 克
水	100 克

做法

水、冷凍柚子汁、砂糖煮滾,放涼待用。

裝組

1. 在已包保鮮膜、直徑 18 厘米的慕斯圈內,擠一
   圈朱古力慕絲,放入黑朱古力脆脆餅底。(看圖
   20-21)

2. 加上一層朱古力手指餅,塗上柚子糖漿。(看圖
   22)

3. 擠上一層朱古力慕絲,放入已裁約直徑 18 厘米
   的柚子餡心。(看圖 23)

4. 把餘下的朱古力慕絲填滿餅模，放入冰櫃冷凍至結冰。（看圖 24）

5. 將餅移離餅模，放在網架上，在餅面邊緣澆一周黑朱古力淋面。（看圖 25）

6. 淋的時候先繞邊緣澆一周，確定邊緣全部沾到，再在中央一氣呵成大量淋下，讓淋面自然流下，覆蓋全個餅，然後用兩手搖一搖網架，讓多餘的淋面滴落，略放一會，才用拉刀移走，飾上裝飾。(看圖 26-27)

** 可做 1 個 18 厘米的蛋糕

貼士

1 日本愛媛縣柚子汁和柚子皮具有攝人的芬芳。在大型超市有該新鮮柚子出售，可惜品質和供應期都不穩定。幸好，近期香港引入了冷凍的柚子製品，讓大家四時都能享受這份芳香。

2 如用大塊朱古力，宜先把朱古力切碎才熱溶。隔熱水溶朱古力時，勿讓朱古力接觸水分，否則會結塊。另外，亦可用微波爐以中大火力慢慢地以每次十多秒溶解朱古力，每次取出攪拌，直至完全溶解，但切勿一次性地長時間以微波加熱，否則朱古力會燒焦。

3 朱古力淋面令甜點賣相更高貴，但通常只會用於冷凍甜點表面，不會淋在海綿蛋糕或其他吸水或表面不平的糕點上。淋面要薄而平均、光亮平滑、沒有氣泡和細膩無暇，這並不簡單。首先要採用較深色的可可粉，蛋糕冷凍後收縮，使中央凹陷，中央便會過厚，一些專業餅師會預留一些慕絲來填補，冷藏結冰後再淋面。淋面時蛋糕必須結冰，朱古力淋面約30℃，此兩條件缺一不可，否則淋面會過薄而看到蛋糕體，或凝固得太快，使表面不平滑。

Dark Chocolate Mousse with Yuzu Jelly Filling

Production workflow

1. Make the Yuzu filling.
2. Make the chocolate ladyfinger.
3. Make the chocolate crispy base.
4. Make the dark chocolate mousse.
5. Make the dark chocolate glaze.
6. Make the Yuzu syrup.
7. Assemble the cake.

Yuzu Filling

Ingredients

40 g frozen Yuzu juice ❶

35 g water

50 g sugar

4 g gelatine leaf (soaked in cold water until soft; drained)

20 g frozen Yuzu zest

Method

1. Bring water and sugar to the boil. Remove from heat. Then add Yuzu juice, frozen Yuzu zest and gelatine leaf. Mix well. (see pictures 1-2)
2. Pour the mixture into an 15-cm round mould. Freeze it. (see pictures 3)
** Makes a 1 cm x 15 cm Yuze filling

Chocolate Ladyfinger

50 g bread flour

15 g cocoa powder

60 g egg yolk

30 g sugar (A)

100 g egg whites

30 g sugar (B)

40 g cornflour

30 g sugar (C)

Method

1. Refer to Coffee Mousse Charlotte on p.47 for step 1 to 4.
2. Transfer the batter into a piping bag. Pipe a round patty 18 cm in diameter on a baking tray lined with non-stick baking lining or silicone mat. Sift icing sugar on top. (see pictures 4-6)
3. Bake in a preheated oven at 170°C for about 15 minutes.

Chocolate Crispy Base

Ingredients

80 g crunchy wheat flakes (pailleté feuilletine)

50 g dark chocolate (over 50% cocoa)

50 g hazelnut paste

Method

1. Melt the chocolate over a hot water bath or in a microwave up to 50°C.❷ (see pictures 7-8)
2. Add hazelnut paste and mix well. Stir in crunchy wheat flakes. (see picture 9)
3. Press onto the bottom of a 15-cm round cake mould. Refrigerate until set. (see picture 10)

Dark Chocolate Mousse

Ingredients

160 g dark chocolate (over 50% cocoa; chopped)

100 g whipping cream

100 g milk

20 g sugar

40 g egg yolk

250 g whipped cream

5 g gelatine leaf (soaked in cold water until soft; drained)

Method

1. Mix egg yolk with sugar.
2. Bring milk and whipping cream to the boil in a pot.
3. Pour the hot cream mixture from step 2 into the egg yolk mixture from step 1. Whisk well. Pass it through a sieve. Then cook in a sauce pan to 85°C. (see pictures 11)
4. Sift the mixture onto the chopped chocolate. Stir until chocolate melts and add soaked gelatine leaf. Stir until gelatine dissolves. Leave it to cool completely. (see pictures 12-13)
5. Fold in whipped cream. (see pictures 14-15)

Dark Chocolate Glaze ❸

Ingredients

150 g water

120 g sugar

120 g whipping cream

10 g gelatine leaf (soaked in cold water until soft; drained)

60 g cocoa powder

Method

1. Bring water, whipping cream and sugar to the boil. Add cocoa powder. Cook until it boils again. (see pictures 16-17)
2. Add gelatine leaf. Stir until it dissolves. Sift and let it cool to 30°C. (see pictures 18-19)

Yuzu Syrup

Ingredients

20 g frozen Yuzu juice

60 g sugar

100 g water

Method

Bring all the ingredients to the boil. Leave it to cool and set aside.

Assembly

1. Wrap the bottom of an 18-cm mousse ring in cling film. Pipe a layer of dark chocolate mousse. Place the frozen chocolate crispy centre over the mousse. (see pictures 20-21)
2. Then put the patty of chocolate ladyfinger on top. Brush Yuzu syrup over it. (see picture 22)
3. Pipe another layer of dark chocolate mousse over the ladyfinger patty. Top with the frozen Yuzu filling. (see picture 23)
4. Fill the mousse ring with the remaining dark chocolate mousse. Keep in freezer until frozen. (see picture 24)
5. Unmould the mousse cake. Transfer the cake onto a wire rack over the counter. Pour dark chocolate glaze around the rim of the cake to cover the vertical sides first. (see picture 25)
6. Then pour the remaining glaze at the centre of the cake all at once. Let the glaze run over the whole cake and let the excess drip off. Leave it briefly and use palette knives to transfer the cake to cake stand or serving plate. Garnish. Serve. (see pictures 26-27)

** Makes one 18-cm round cake

Tips

1 Yuzu is a Japanese citrus fruit and those from Ehime Prefecture, Japan are best known for their breathtaking fragrance in both their juice and zest. Major supermarkets in Hong Kong do carry fresh Yuzu from time to time but with varying freshness and quality. Luckily, frozen Yuzu products are also available lately, so that its breathtaking fragrance is also made available all year round.

2 If you're using chocolate bars or slabs, always chop them up before melting them. When you heat chocolate over a hot water bath or in a double boiler, do not splash any water onto the chocolate. Otherwise, it will separate and turn lumpy. If you're melting chocolate in a microwave, heat it up for 10 seconds at a time. Stir well after each heating session until it melts completely. Do not microwave in one go to melt it completely. As microwave heats food up unevenly, the chocolate along the rim of the container will burn before that at the centre is even melted.

3 The chocolate glaze makes the cake more elegant and sumptuous. But it is usually applied on frozen dessert only, but not on sponge cake or other cakes with an absorbent or uneven surface. The perfect glaze should be even and thin; shiny and smooth; free of blister or bubble; delicate and flawless. First of all, always use a dark cocoa powder for the glaze. Secondly, sometimes a cake may shrink more at the centre than on the rim after frozen. If you pour the glaze on it straight, the glaze will be too thick at the centre. Professional pastry chefs will patch the sunken centre with extra mousse first. Then freeze it again before pouring the glaze on it. Finally, there are two conditions that have to be fulfilled for a glaze to be perfect: that the cake has to be frozen; and that the glaze is right at 30˚C. If the glaze is too hot or the cake is not cold enough, the glaze will be too thin and the cake shows through the glaze. If the glaze is not hot enough, the glaze will set too quickly and become uneven.

我深信"Making dessert is easy, with direction"。
看畢這本甜品書，希望大家在烘焙道上不用再時刻如履薄
冰，讓越來越多甜品愛好者如我，在家中廚房也能做出媲
美專業的甜品。

鳴 謝

師友：

Mr. Johnny CHAN　陳景祥先生

　　香港烘焙專業協會副會長，是我最尊敬，對我影響最深遠的師友，多謝他的教導，令我接觸和認識更多烘焙上的知識，擴闊我的領域。

Mr. Daniel LAW　羅洪基先生

　　日航酒店糕餅主廚，與他相識五、六年，和他很投緣，兩次與他一起跳出香港往外地見識和工作。雖然不是常常見面，但親切和善的他總給我不少啟蒙，感謝他為我撰寫序言。

Mr. Ronny LEUNG　梁樂生先生

　　香港會糕餅主廚，初認識他是在《今日烘焙》雜誌，覺得他為人很低調，不苟言笑，後來與他相識，發覺他很喜歡幫助人和提點後輩。認識數年雖不熟稔，但每次見面都有製作甜品的交流和新資訊帶給我，感謝他為我撰寫序言。

Mr. Titan TSANG　曾智波先生

　　香港深灣遊艇會遊艇俱樂部餅師。認識Titan是很偶然的事，是他先在我的blog內留言。雖然認識他不久，但總像多年的老朋友，可能就是大家志趣相投吧。他是獨當一面的糕餅師傅，但藍色大門的大小聚會他也會參加，還分享他的心得，可見他為人相當隨和；還不止，每當他去旅行也帶來當地手信給獨角仙，這令獨角仙深受感動呢，感謝他為我撰寫序言。

藍色大門好友：

Mr. CHAN Ming-ho, 蛙蛙會長

Ms. Erica LAI, 喪e

Ms. Sammy HA, 豬小姐

Ms. Tracy TO, 編輯大人

Ms. May MOK, 陳may

Ms. Minnie KAN, Minnie

Ms. Tendy YIM, Tendy

Ms. Vicky KWAN, 查查

Ms. Michelle Nam, 藍藍

多謝他們騰出寶貴時間、無價友誼來協助我籌備這本書。

特別鳴謝：

攝影師Mr YAN Kin-wai, 威哥和Mr. CHIU Yuk-shing, 屎屎。

綠萼, Mrs. Cass TUNG為書做校對、為文字潤飾和撰寫序言。

支持我的丈夫（大師）、Ms Rachel YAU 和 Ms. Anne CHAN為大家預備膳食，讓我們在製作期間營遍中、西、意、法、日式美食。

Mr. YAM Lok-hin, Hynn 的支持和教導。

西貢大涌口村佳記花園借出拍攝道具

香脆烘焙

豈能戒甜 Not quitting sweets!

作者	Author
獨角仙@藍色大門	Kin
策劃/編輯	Project Editor
	Catherine Tam
攝影	Photographer
	Wai & Mr. C Photography
美術統籌及設計	Art Direction & Design
	Ame
美術設計	Design
	Man Lo

出版者 Publisher

Forms Kitchen Publishing Co.,
an imprint of Forms Publications (HK) Co. Ltd.
香港筲箕灣耀興道3號東滙廣場9樓 9/F., Eastern Central Plaza, 3 Yiu Hing Road,
Shau Kei Wan, Hong Kong
電話 Tel: 2976 6570
傳真 Fax: 2597 4003
網址 Web Site: http://www.formspub.com

發行者 Distributor

香港聯合書刊物流有限公司 SUP Publishing Logistics (HK) Ltd.
香港新界大埔汀麗路36號 3/F., C&C Building, 36 Ting Lai Road,
中華商務印刷大廈3字樓 Tai Po, N.T., Hong Kong
電話 Tel: 2150 2100
傳真 Fax: 2407 3062
電郵 Email: info@suplogistics.com.hk

承印者 Printer

合群(中國)印刷包裝有限公司 Powerful (China) Printing & Packing Co., Ltd.

出版日期 Publishing Date
二〇一〇年七月第一次印刷 First print in July 2010

版權所有・不准翻印 All rights reserved.
Copyright ©2010 Forms Publications (HK) Co. Ltd.

ISBN 978-988-18764-7-8
Published in Hong Kong